AF424543

QUANTUM CREATION

MERGING NEVILLE GODDARD'S INSIGHTS WITH MODERN PHYSICS

K. JAYANTH MURALI

Notion Press

QUANTUM CREATION

MERGING NEVILLE GODDARD'S INSIGHTS WITH MODERN PHYSICS

K. JAYANTH MURALI

INDIA • SINGAPORE • MALAYSIA

Copyright © K. Jayanth Murali 2025
All Rights Reserved.

ISBN

Hardcase 979-8-89699-948-5
Paperback 979-8-89699-947-8

This book has been published with all efforts taken to make the material error-free after the consent of the author. However, the author and the publisher do not assume and hereby disclaim any liability to any party for any loss, damage, or disruption caused by errors or omissions, whether such errors or omissions result from negligence, accident, or any other cause.

While every effort has been made to avoid any mistake or omission, this publication is being sold on the condition and understanding that neither the author nor the publishers or printers would be liable in any manner to any person by reason of any mistake or omission in this publication or for any action taken or omitted to be taken or advice rendered or accepted on the basis of this work. For any defect in printing or binding the publishers will be liable only to replace the defective copy by another copy of this work
then available.

Dedication

To the invisible threads of the quantum realm,
where dreams take shape, and thoughts unfold.
This is for those who whisper to the stars
and dare to call the unknown their home.
May your journey into this wondrous web of possibility
ignite the spark that creates your universe.

Praise for Quantum Creation

"Combining elegance of quantum science with the profundity of spiritual insight, this book is an essential read for anyone who dares to dream and create."

Sandeep Rai Rathore, IPS,
Director General of Police, (Training), Tamil Nadu.

"This book is a rare and powerful fusion of logic and wonder. Dr. Murali takes readers on a journey that is both deeply personal and universally relevant. Through *Quantum Creation,* we discover that the principles of quantum physics are not confined to laboratories—they are the very fabric of our lives. Neville Goddard's teachings come alive in these pages, offering hope, clarity, and actionable steps for transforming our reality. An extraordinary achievement!"

Yogesh Singh
Principal Chief Conservator of Forests (retd.)

"Policing requires resilience and the ability to lead under pressure. This book teaches how to align our vision with outcomes using the principles of quantum mechanics. It's a game-changer for anyone seeking clarity and focus in high-stakes careers."

Mahesh Aggarwal, IPS
Addl. DGP, Armed Police, Tamilnadu.

"*Quantum Creation* beautifully intertwines the timeless wisdom of Neville Goddard with the groundbreaking discoveries of quantum physics. Dr. Murali's work is a beacon for those seeking to harness the power of their minds to shape their destiny."

N. Venu Gopal, IPS,
Additional Director, CBI.

"As a police officer, I've faced challenges that seemed insurmountable, but this book gave me a fresh perspective. The connection between belief and outcome is beautifully explained with practical steps for application. A must-read for leaders and professionals in service."

Dr J Loganathan, IPS
IGP/Commissioner of Police, Madurai City

"This book attempts to find common ground between the Bodhi tree and the falling of the apple."

Sanjay Pinto
Advocate, Columnist & Author, Former Resident Editor – NDTV 24x7

"In *Quantum Creation*, Dr. Murali achieves something remarkable: he makes the complex principles of quantum mechanics accessible while intertwining them with the profound teachings of Neville Goddard. The result is a book that is as enlightening as it is empowering. It's a guide for the modern creator—a reminder that we are all architects of our own realities. The depth of research combined with the warmth of storytelling makes this a truly transformative read."

Dr K S Vijay Elangova
Author and Senior Journalist

"*Quantum Creation* is a groundbreaking work that beautifully aligns with the philosophy of healing. As a kidney specialist working with the TANKER Foundation, I've seen the profound impact of belief and hope on recovery. Just as kidneys filter and purify the blood, this book helps purify the mind of limiting beliefs, enabling readers to align with their highest potential. It's an inspiring guide for creating a life of health and abundance."

Dr. Georgi Abraham
Founder Trustee, TANKER Foundation

"In the operating room, I see life at its most fragile and most resilient. "*Quantum Creation*" captures this duality beautifully, showing how belief and focused intention can create profound change. The liver, with its remarkable regenerative capacity, is a testament to the principles outlined in this book—where healing is not just physical but begins in the mind. This is a powerful guide for anyone who wants to take charge of their destiny and health."

Dr. Karthik Mathivanan D.N.B
Programme Director, Institute of Liver Disease and Multiorgan Transplantation MIOT Hospitals Private Limited.

"I didn't think quantum mechanics could have practical applications in daily life until I read this book. Now, I use its principles to guide my business decisions."

Dr. Sam Paul
Director & Legal Head
Paulsons Beauty & Fashion Pvt. Ltd./Pauls System Technologies Pvt. Ltd.

"*Quantum Creation* is a profound exploration of the limitless potential of the human mind, blending modern physics with timeless wisdom. As a police officer, I've witnessed firsthand how focus, belief, and resilience can shape outcomes in the most challenging situations. This book provides the tools to harness those principles, not just in law enforcement but in every aspect of life. A must-read for those ready to take charge of their reality."

A. Pavan Kumar Reddy, IPS
Deputy Commissioner of Police, Tambaram, TN

"*Quantum Creation* is a masterpiece for dreamers and doers alike."

G.Stalin
Superintendent of Police, Mayiladuthurai District

"Dr. K. Jayanth Murali's "*Quantum Creation*" is a brilliant fusion of science and spirit, showing how quantum principles can spark creativity and flow in writing and podcasting. By mastering these insights, we can align with the frequencies of inspiration and produce work that profoundly impacts our audience. A must-read for storytellers and creators."

Vamsi Deepak Sankar
President – Professional Speakers Association of India, Chennai Chapter

---

"In the courtroom, logic and reason dominate. Yet, *Quantum Creation* showed me how our subconscious mind and the quantum world influence outcomes more than we realize. Truly enlightening."

Sanjeev Yadav
Advocate, New Delhi

Previously Published Books by the Author

1. **42 Mondays:**
   On Emerging Technologies for Policing

2. **Soliloquies on Future Policing:**
   An Anthology on Emerging Technologies, Cybersecurity and Law Enforcement

3. **Enkindling the Endorphins of Endurance:**
   Your Ultimate Go – To Guide for Running Your First 5K, 10K, Half and Full Marathon

4. **A Random Potpourri:**
   An Anthology of Remnant Scrawls From a Policeman's Diary

5. **The Art of Peak Performance:**
   Hacking the Body and Mind For Peak Success

6. **Marathon (*Tamil*)**

Contents

PART 1: FOUNDATIONS AND PRINCIPLES OF QUANTUM MANIFESTATION1

PART 2: TOOLS, APPLICATIONS, AND MASTERY
OF QUANTUM CREATION

Contents

Foreword

SRINIVAS R. REDDY, I.F.S.,
PRINCIPAL CHIEF CONSERVATOR OF FORESTS
(HEAD OF FOREST FORCE)

Forest Headquarters,
Velachery Main Road,
Guindy, Chennai - 600 032.
Tele No. 044-24348059
Email: pccfhod@gmail.com
tnforest@nic.in

Dated: 04.12.2024

Nestled amidst the rolling hills of Ooty and Coonoor, where misty mornings embrace lush green canopies, and the whisper of the wind carries the timeless wisdom of ancient forests, one learns to appreciate the delicate balance of existence. As a forester, I have spent decades immersed in the Nilgiris, in the harmony of nature, observing the intricate interplay between every tree, stream, and creature. This balance, which is both fragile and resilient, is a microcosm of a grander truth—a truth beautifully captured in Dr. K. Jayanth Murali's " Quantum Creation: Merging Neville Goddard's Insights with Modern Physics".

Just as the forests of Tamil Nadu speak a language of interconnectedness, Dr Murali unveils the profound web of quantum mechanics, revealing that our thoughts and intentions ripple through the cosmos, shaping reality much like how a single seed can birth a forest. This book bridges two worlds: the tangible, natural realm I've spent my life protecting and the intangible quantum universe, where possibilities shimmer like the dewdrops on Nilgiri leaves.

The Symphony of Creation

In these pages, Dr Murali invites us to embark on a journey that mirrors the rhythm of the forest—a journey of growth, resilience, and renewal. Like the banyan tree, whose roots reach deep into the earth while its branches stretch toward the sky, this book connects the grounded teachings of Neville Goddard with the lofty principles of quantum physics. It is a guide for those seeking not just to live but to thrive to understand their role as both caretakers and creators of their world.

Dr Murali's exploration of quantum phenomena, such as the Observer Effect and quantum entanglement, resonates deeply with my experience in the wild. In the forests, every action—no matter how small—leaves an imprint, just as every thought and intention leaves an imprint on the quantum field. His writing reveals a truth that foresters have long known: we are never separate from the ecosystems we inhabit, whether physical or metaphysical.

A Map for the Modern Explorer

What sets *Quantum Creation* apart is its practical wisdom. Much like the field guides that help us navigate the dense jungles of Mudumalai or the shola forests of the Nilgiris, this book offers a "Quantum Toolbox" to navigate the complexities of life. Visualisation, affirmations, and meditation are presented as abstract concepts and tools to reclaim our innate power to shape our destiny.

Dr Murali's anecdotes, interwoven with the narrative, bring authenticity and inspiration. As I read, I could almost hear the rustle of Nilgiri leaves and the call of the Malabar whistling thrush—reminders that, like the forest, our lives are ever-changing and full of potential.

Bridging Worlds

To me, this book is a bridge—a way to connect the tangible lessons of nature with the abstract beauty of quantum theory. It reminds us that, whether tending to a sapling or nurturing a dream, the principles of creation remain the same. The forest teaches patience and intention, and so does quantum physics.

Dr Murali's poetic prose flows as effortlessly as the streams of Coonoor, drawing readers into a dance of energy, possibility, and creation. His work transcends the boundaries of science and spirituality, offering readers a glimpse of a universe that is both familiar and profound.

A Beacon of Hope

In a world increasingly disconnected from nature and from itself, *Quantum Creation* is a beacon of hope. It reminds us that we are not merely observers but active participants in the grand dance of life. Whether you are a scientist, a spiritual seeker, or someone looking to align with the rhythms of existence, this book will transform you.

As someone who has spent a lifetime among the forests of Tamil Nadu, I see this book as a celebration of the interconnectedness that defines our world. It is a testament to the power of intention, the beauty of creation, and the infinite potential within each of us.

To those who hold this book, I offer these words: Let its wisdom guide you as the forests have guided me. Savour its lessons, act upon its insights and watch your life transform into something extraordinary.

Much like a forest nurtures life, Dr. Murali has given us a masterpiece that nurtures the soul. It is an honour to pen this foreword and to witness the profound impact this work will have on all who read it.

With heartfelt gratitude,

Srinivas R. Reddy, IFS.,
Principal Chief Conservator of Forests
(Head of Forest Force)
Tamil Nadu

Acknowledgments

In the vast quantum field where thoughts ripple into reality and possibilities collapse into existence, this book, Quantum Creation: Merging Neville Goddard's Insights with Modern Physics, is but a wave made manifest by the entanglement of countless luminous minds and loving souls. To each of you who exist within this shared superposition of creativity and purpose, my gratitude transcends time and space, echoing in the infinite.

First and foremost, I honor the cosmic force that is my wife, Jayanthi Murali, Chairperson, Tamilnadu Pollution Control Board. Her unwavering presence is the observer that brings order to the chaos of my creative mind, collapsing uncertainty into profound clarity. Like the Higgs field lending mass to matter, her grace and patience have given this endeavor its substance. Every quantum leap within these pages is intertwined with the resonance of her love.

To my parents, the origin of my wave function, whose timeless wisdom and grounding energy set the parameters of my being. They are the gravitational anchors in the swirling galaxies of my life. From their nurturing, I derived the constants of resilience, integrity, and wonder that shaped the foundation of this exploration. I remain an eternal particle in their boundless field of love.

To my daughters, Tanya and Sonya, the dual photons of my existence, whose light splits into infinite inspiration. Their laughter, their faith, and their encouragement are quantum probabilities realized in every decision to keep moving forward. In the entanglement of our shared lives, they remain my greatest treasures, illuminating the path to endless possibilities.

To my siblings, Ashok and Priya, you are the harmonic oscillations in my quantum state, always in phase with my journey. Your belief in me resonates like constructive interference, amplifying every effort, every dream. I am grateful for the coherence your presence brings to my life's waveform.

To Vidhya, whose precision and dedication are the Planck constants of this book's creation. Your meticulous attention has been the fine-tuning that transforms noise into signal, chaos into order. In the quantum dance of creativity and execution, your contributions are the binding force of this reality.

To my steadfast friends, Rafiq Bhaiya and Bhaskar, you are the dark energy that expands my universe, driving me forward even when the path seemed unclear. Rafiq, your steady presence is the lighthouse guiding me through uncertain waters. Bhaskar, your belief in my potential is the quantum nudge that collapses doubt into determination. You both are my constants in an ever-evolving equation.

To the team at Notion Press—Ms. Vandana Valsakumar, Mr. Akash, Ms. Nidhi Shetty, Ms. Anjani Ram and their leader, Mr. Naveen—your vision and precision are akin to the quantum entanglement that binds all aspects of this book into a seamless whole. Your collective energy transformed an abstract idea into a physical reality, for which I am infinitely grateful.

To my trusted companions—Murali, Govindan, Suresh, Thiyagarajan, and Arun Prasad—your tireless support is the scaffolding upon which this creation stands. Like unseen quantum particles, your contributions may go unnoticed by the larger world, but they are foundational to the existence of this work.

To the hidden catalysts, the innumerable souls whose words, gestures, and energy have influenced this narrative—your presence, whether visible or veiled, is interwoven into the spacetime fabric of this book. Every thought, every idea, carries your imprint.

Finally, to you, the reader, the ultimate observer in this quantum experiment. By opening these pages, you collapse the wave of potential into a lived experience. You complete the entangled circuit, turning this book into a shared reality. May these words inspire you to see the world not just as it is, but as it can be—a boundless quantum field of possibility waiting to be shaped by your intention.

Together, we co-create. Together, we transcend.

Preface: The Quantum Journey

"Imagination is the key to unlocking the wonders of life. It allows us to glimpse the incredible things that lie ahead."

"Imagination is more important than knowledge."

– Albert Einstein

In the heart of a quaint town, nestled within the tranquil sanctuary of my childhood home, I often found myself lost in a boundless sea of imagination. These were not mere fleeting fancies of a boy but vivid, vibrant visions of a future brimming with purpose and success. I saw myself, resolute and dignified, adorned in the crisp uniform of a police officer, a sense of honour and responsibility swelling within my heart. This vision, a beacon of hope, illuminated my path, guiding me through life's inevitable twists and turns.

Looking back, I realise this book, Quantum Creation: Merging Neville Goddard's Insights with Modern Physics, transcends the realm of mere ideas and hypotheses. It stands as a testament to the profound power of imagination and the transformative influence of quantum physics. Within these pages, I will unveil the profound impact these concepts have had on my life, revealing how I transformed my wildest dreams into tangible reality. I hope that this journey will inspire you to unlock your inner power.

The Search for Purpose

Amidst the pursuit of a PhD in microbiology in Delhi, a profound sense of unease began to gnaw at me. A strange void, a gaping chasm, seemed to exist within my soul. Despite my academic achievements, a deep-seated discontentment lingered, a nagging feeling that I was not fully utilising my potential and not making a significant impact on the world.

In this period of introspection, a newspaper advertisement for a book on self-hypnosis by Pradeep Aggarwal caught my eye. Before the ubiquitous presence of mobile phones and the internet, it was the analogue era, a world where information flowed through ink and paper. Intrigued by the promise of personal transformation, I dispatched a postal order of Rs 50/- and eagerly awaited the book's arrival. Twenty days later, the familiar figure of our local postman delivered the package, a small beacon of hope amidst the mundane routine of daily life.

The book, though simple in its presentation, held a profound power. It contained exercises in auto-suggestion and self-hypnosis, offering tools to unlock the mind's untapped potential. I immersed myself in its pages, diligently practising the techniques, visualising myself as a police officer, embodying the integrity and dedication that role demanded. At that time, I was unaware of the profound connection between these exercises and the principles of quantum physics. I only knew that they resonated deeply within me, aligning perfectly with my inner world.

Gradually, the seeds of my dream began to sprout. With each visualisation, I felt myself drawing closer to my goal. The crispness of the uniform, the weight of responsibility, the satisfaction of serving and protecting – these sensations, once mere figments of imagination, began to feel increasingly accurate. And, as fate would have it, I cleared the UPSC examination, fulfilling my lifelong aspiration of joining the Indian Police Service.

Years later, through deep introspection and ongoing exploration, I began to understand the profound impact of those initial exercises. My success was not merely a product of hard work and perseverance; it manifested a vision I had nurtured in my imagination long before it became a reality.

The Observer Effect: Manifesting Reality

One of the fundamental tenets of quantum mechanics is the observer effect, which posits that observation can influence an event's outcome. In manifestation, our thoughts and intentions wield immense power in shaping our reality. When we set our sights on a specific goal and wholeheartedly believe in its attainment, we transform the realm of infinite possibilities into a concrete reality.

Throughout my journey towards becoming a police officer, I embraced this principle with unwavering dedication. I visualised myself in uniform, carrying out my duties with confidence and integrity. Through meditation and self-hypnosis, I immersed myself in the feeling of having already achieved my goal. Visualising success bolstered my determination and harmonised my energy with the vibration of my desired outcome. From the quantum physics perspective, my unwavering observation acted as a catalyst, manifesting my dream into reality.

In 1993, while stationed in Salem, amidst the tense backdrop of the infamous Veerappan operations, I faced a stark reminder of the fragility of life. The Palar blast, a tragic event etched in my memory, shook the very foundations of the region. On April 9th, a bus carrying members of the Special Task Force was ambushed near Surakkamaduvu. Veerappan, the notorious bandit, detonated landmines, resulting in a devastating explosion that claimed the lives of 22 people, including police officers, forest officials, and innocent civilians. Thirteen others were critically injured. The chaos and despair that ensued were palpable, even miles away.

I was assigned to Gokulam Hospital to oversee security arrangements as the wounded were rushed for treatment. The atmosphere was tense, further amplified by the impending visit of the Honorable Chief Minister, J. Jayalalithaa, who planned to meet with the survivors. For two days, I remained vigilant, ensuring the security arrangements were impeccable.

During this period, Mr Vijay Kumar, DIG, SSG, and Ms Agarwal, the Joint Superintendent of Police, visited the hospital to assess the situation and offer their condolences to the survivors. While they commended my dedication, I couldn't ignore the subtle disappointment in their eyes as they glanced at my physique. Their unspoken judgment, though slight, resonated deeply within me.

That night, their silent critique echoed in my mind like a haunting melody. It wasn't just about my weight but about who I had become – lethargic, out of shape, a shadow of the man I once aspired to be.

Amidst the backdrop of tragedy and self-reflection, a new resolve began to stir within me. The unspoken criticism, coupled with the weight of the situation, served as a powerful catalyst for introspection and change. I reached for an old companion, a book I had almost forgotten: Pradeep Aggarwal's work on self-hypnosis and auto-suggestion. I read it years ago, and its teachings ignited a spark. Now, I returned to its pages, seeking solace and guidance.

I began to visualise myself as a fitter, healthier version of myself. This was not wishful thinking; I saw, felt, and believed in this transformation. I imagined myself running through the outskirts of Salem, the wind whipping through my hair, a surge of energy coursing through my veins. I translated these visualisations into action, beginning with small steps – short runs within the District Armed Reserve police grounds, each step a small victory against the inertia that had crept into my life. At that time, I could not have imagined that I would one day become a marathon runner, breaking records and pushing the boundaries of human endurance. My focus was singular: to reclaim my health, one step at a time.

By 1995, I had been promoted to Superintendent of Police, Tanjore, and had regained much of my fitness. However, another insidious habit continued to cling to me – smoking. Despite my best efforts, I could not break free from its grip.

My subconscious mind began to rebel. In my dreams, my father appeared, his face etched with disapproval. My mother, too, visited me in these dreamscapes, her eyes filled with concern and disappointment. I knew I had to quit. I resumed my practice of auto-suggestion and self-hypnosis, vividly visualising myself as a non-smoker. I could almost taste the freshness of a smoke-free life, the promised freedom. Yet, the grip of addiction remained strong.

Then, one evening, everything changed.

I had returned home after a long day, eager to embrace my elder daughter, Anisha. As I leaned in to hug her, she recoiled slightly. "Dad," she said, her voice soft yet firm, "please don't come near me. I don't like the smell of nicotine on your breath."

Her words struck me like a bolt of lightning. In that instant, all my visualisations, every moment of self-hypnosis, and every intention I had ever set seemed to converge, exploding within me like a supernova. Years of preparation and cultivating the right mindset culminated in this transformative moment.

I didn't just decide to quit smoking; I stopped. There were no cravings, no withdrawal symptoms, and no lingering thoughts of reaching for a cigarette. It was as if a switch had been flipped, releasing me from the shackles of addiction.

Looking back, I realise this was not merely an act of willpower. It manifested quantum principles in action – a symphony of intention, assumption, and entanglement. My daughter's words served as the catalyst, but the groundwork had been laid long before. The dreams of my parents, the countless hours spent in self-hypnosis, the unwavering power of my visualisations – all these elements converged to create a reality where I was free.

This experience taught me a profound lesson: transformation is not a singular event but a symphony of intentions, beliefs, and actions. The quiet work we do within ourselves, the seeds we plant in our subconscious, creates the fertile ground for the miraculous to blossom. When we align ourselves with quantum principles, we unlock our true potential and manifest our desired life.

Superposition and Possibility: Expanding Potential

In the fascinating realm of quantum mechanics, particles can exist in multiple states simultaneously until observed – a phenomenon known as superposition. This principle mirrors the boundless potential of human existence. When we embrace the possibility of various paths, the universe guides us towards the most fulfilling outcome.

As an author, I have experienced the beauty of superposition firsthand. My writing journey has never been linear, confined to a single topic or genre. Instead, I embraced the freedom to explore a spectrum of ideas, each unique and compelling.

My literary journey began with technology, delving into the transformative potential of emerging innovations. 42 Mondays and Soliloquies on Future Policing reflect my fascination with technology's role in law enforcement and cybersecurity, envisioning a future where innovation enhances public safety.

My writing ventured into the world of endurance from the realm of technology. In Enkindling the Endorphins of Endurance, I captured the grit, resilience, and unwavering spirit of marathon runners. The Art of Peak Performance followed, exploring the fascinating science of biohacking and the art of optimising body and mind for peak performance.

However, as I have learned, life is a tapestry woven with threads of diverse experiences. In A Random Potpourri, I allowed my pen to wander freely, exploring reflections on life, health, desires, and the fleeting beauty of everyday moments.

And now, with Quantum Creation: Merging Neville Goddard's Insights with Modern Physics, I find myself weaving together the threads of science and spirituality. This book is a culmination of ideas that have shaped my life – an exploration of how imagination and quantum principles can transform dreams into reality.

Each book I have written serves as a testament to the power of superposition – a willingness to embrace the unknown, to allow creativity to flow freely, unhindered by limitations. By adopting the possibility of multiple outcomes, I have traversed diverse terrains, from the corridors of technology to the trails of endurance, the wonders of biohacking, and now, the profound depths of quantum manifestation. Writing has expanded my horizons, reminding me that we create space for the extraordinary when we surrender to the unknown.

Quantum Entanglement and Connectedness

Quantum entanglement, one of the most captivating principles of quantum mechanics, reveals the profound interconnectedness of all existence. It demonstrates that the state of one particle can instantaneously influence another, regardless of the distance separating them. At its core, this principle reminds us that nothing exists in isolation – our thoughts, emotions, and actions create ripples that extend far beyond our immediate sphere of influence.

In my journey as a marathon runner, I have experienced this interconnectedness firsthand. Running, at its essence, is more than a solitary pursuit; it is a testament to the interplay between body and mind, a dance between solitude and the shared strength of a community. In the camaraderie of fellow runners, I found a unique form of quantum entanglement – a silent understanding that we were all striving towards something greater than ourselves. Their stories, their struggles, and their triumphs became my inspiration.

Every record I broke, including my two entries in the Asia Book of Records, bore the unseen fingerprints of those who ran beside me, their energy intertwined with mine.

This interconnectedness extends far beyond the realm of running. During my tenure as a police officer, I witnessed firsthand how the actions of a single individual can create profound ripples across entire organisations and communities. A single idea, driven by conviction and intent, can transform the trajectory of countless lives. Whether it was recovering stolen artefacts or exposing systemic corruption, the efforts of a few inspired many, creating a network of change that felt as instantaneous and profound as quantum entanglement itself.

Quantum entanglement is more than just a scientific principle; it is a profound reminder of the invisible threads that bind us to one another and the universe. Our thoughts, intentions, and actions create ripples, influencing lives and shaping outcomes in ways we may never fully comprehend. The true power of entanglement lies in this understanding: every choice we make, every intention we set, carries the potential to create waves of change. When we act with awareness and purpose, we participate actively in this beautiful, interconnected web of existence.

Neville Goddard's Teachings: The Power of Imagination

Neville Goddard's teachings on the power of imagination have played a pivotal role in shaping my journey. He believed profoundly in the ability of imagination to shape our reality. When we embrace the feeling of our wish fulfilled, we align our inner state with the desired outcome, making its manifestation inevitable.

Throughout my life, I have consistently relied on the magic of imagination to bring my dreams to life. Imagining myself as a successful police officer, a renowned author, or a record-breaking marathon runner, I fully embraced the emotions associated with these achievements. This practice bolstered my self-confidence and set in motion the events and circumstances that ultimately led to their realisation.

However, the power of imagination extends far beyond personal gain. It opens a gateway to a deeper understanding of the cosmos and our place within it. According to Goddard, the world acts as a mirror, reflecting our inner thoughts and actions. By harnessing the power of imagination, we have the potential to transform our own lives and influence the world around us.

Practical Techniques for Quantum Manifestation

In the following chapters, you will discover practical techniques for incorporating quantum principles into your manifestation practices. By integrating various methods, such as visualisation, affirmations, meditation, and intention setting, you can tap into the boundless energy of the quantum field and manifest the life you truly desire.

One effective method is to elevate your energy by engaging in activities that uplift your spirit. Incorporating exercise, meditation, and surrounding yourself with positive influences can effortlessly elevate your vibrational frequency, attracting more positive experiences into your life.

Another crucial technique is to establish strong intentions. When you possess a clear vision and radiate positive energy, you naturally attract what you desire. You can unlock your true potential by embracing the power of love and aligning yourself with something greater than yourself.

Overcoming Blocks and Misconceptions

The journey of quantum manifestation is not without its challenges. In the subsequent chapters, we will explore standard blocks and misconceptions that can hinder your progress. We will delve into the importance of overcoming self-doubt, releasing limiting beliefs, and cultivating a mindset of abundance.

I invite you to embark on this transformative journey with me. Let us unlock the power of quantum creation and manifest the life we deserve.

Introduction: Awakening the Quantum Mind

"The mind is everything. What you think, you become."

– Buddha

As the monsoon clouds cleared a bit, casting playful shadows on the café window, I couldn't help but get lost in my own musings on that muggy afternoon in Chennai. My notebook lay open in front of me, filled with scribbles that seemed to capture the chaotic essence of the universe. As I savoured my cup of coffee, a memory resurfaced—one that inspired me to delve into the fascinating connection between Neville Goddard's concepts and the mysteries of quantum physics.

It was not just any ordinary remark from someone I barely knew that ignited this adventure but a captivating story I stumbled upon while delving into the life of Dr. Joe Dispenza, a highly esteemed neuroscientist, author, and speaker. Dispenza's story is truly extraordinary, showcasing the incredible possibilities that emerge when science, consciousness, and human willpower converge.

In 1986, Dr. Dispenza experienced a devastating accident during a triathlon in Palm Springs, resulting in fractures to six vertebrae in his spine. The prognosis was not optimistic. Medical professionals advised a surgical procedure that would fuse his spinal segments, potentially resulting in long-term physical restrictions. However, Dispenza, with a profound grasp of the body's regenerative powers and a firm foundation in quantum physics, made a courageous choice. Instead of undergoing the surgery, he relied on the power of his mind to heal himself. He remained motionless for weeks, envisioning every vertebra aligning perfectly. He spent hours each day in deep meditation, immersing himself in the power of positive visualisation rather than dwelling on his current hardships.

Remarkably, in just nine months, Dr. Dispenza made a complete recovery without the need for any invasive surgery. This experience had a profound impact on him and solidified his faith in the powerful connection between the body's innate wisdom and the conscious mind's capacity to shape our world. It beautifully resonates with Neville Goddard's teachings on the Law of Assumption.

The Intersection of Science and Spirituality

Dr. Dispenza's story goes beyond being a personal triumph; it serves as a compelling example that delves into the ideas this book aims to explore. *Quantum Creation: Merging Neville Goddard's Insights with Modern Physics* explores the fascinating connection between quantum mechanics and Goddard's teachings on manifestation. By delving into the abstract concepts of quantum mechanics, this book aims to enhance our understanding and application of Goddard's profound insights.

This book invites readers to go beyond conventional wisdom and delve into a realm where spirituality and science coexist in harmony. It explores the fascinating connection between our spiritual beliefs and the principles of quantum physics, revealing how they intertwine to shape our reality.

Throughout my career as a police officer, I have come across situations that have left me puzzled and searching for answers. During my time as the Director of Vigilance and Anti-Corruption, I frequently encountered challenging cases involving deeply rooted corruption. One particularly fascinating case involved a complex embezzlement scheme that appeared incredibly difficult to solve. We faced a formidable challenge as our usual methods of investigation proved ineffective against an elusive and mysterious force. During this period, I started implementing the principles of visualisation and focused intent, which resonated with both quantum theory and Goddard's teachings.

Every evening, I would imagine the intricate web of corruption slowly coming undone as each thread revealed another clue. I envisioned the wrongdoers being held accountable and the stolen funds being reclaimed. As I delved deeper into my investigation, fresh evidence surfaced, causing the case to unravel. In just a few months, we dismantled the entire operation completely, resulting in the recovery of lakhs of rupees. This experience further solidified my conviction in the incredible ability of the human mind to shape the world. It's a belief that resonates with the principles of quantum mechanics and the teachings of Goddard.

Quantum mechanics offers a captivating tale of exploration that pushes the boundaries of our perception of reality. Envision a time when science was based on certainty and determinism, and the classical universe of Newtonian physics reigned with authority. However, hidden beneath the apparent simplicity lay a captivating realm—the quantum field—where particles exhibited astonishing behaviours that defied all conventional expectations.

In the early 20th century, there was a remarkable shift in our understanding of the cosmos. Influenced by pioneering physicists such as Niels Bohr, Werner Heisenberg, and Erwin Schrödinger, the captivating realm of quantum mechanics took shape. It opened up a world of endless possibilities that pushed the boundaries of what we thought was real. At the core of this transformation, a groundbreaking concept emerged: wave-particle duality. Albert Einstein initially introduced this concept as he delved into the study of the photoelectric effect. This intriguing phenomenon suggests that particles such as electrons and photons can exist as both waves and particles at the same time. It presents a paradox that defies conventional wisdom yet lies at the very core of quantum theory.

Bridging Worlds: Neville Goddard's Spiritual Realisations

Even before quantum physics became popular, Neville Goddard, a visionary teacher and mystic, possessed a deep understanding of the mind's creative power. According to his teachings, creation is primarily a result of our imagination. By embracing the power of visualisation, we can unlock the boundless potential of the universe and manifest our deepest desires.

Goddard's insights have a deep connection with quantum mechanics, specifically in recognising the significant impact of the observer on shaping reality. While Goddard highlighted the importance of consciousness, quantum mechanics brings up the observer effect, which suggests that observation can influence the results of an experiment. This book aims to connect these worlds and provide readers with a guide to fully harness this potent synergy in their own lives.

Throughout my personal experiences, I have witnessed the principles of Goddard's teachings, and quantum mechanics come to life with tangible results. As an officer, I've encountered situations where the challenges appeared overwhelming. However, by staying focused on the desired outcome, whether it was solving a case or restoring law and order, I saw firsthand how my determination played a vital role in achieving success. It's almost magical how a positive belief can manifest into reality.

One of the most memorable cases I worked on was the recovery of stolen idols from a network of international smugglers. Foreign museums and private collectors had unfortunately taken these priceless idols, with their rich cultural and historical significance, away. The task appeared challenging, with leads spread out all over the world. Through the application of Goddard's principle of assuming the feeling of the wish fulfilled, I vividly imagined the idols being returned to their rightful places. Through deep contemplation and unwavering determination, we ultimately rescued many invaluable artefacts.

This was not only a triumph for law enforcement but also a powerful demonstration of the impact of unwavering determination.

Exploring Fresh Perspectives and Scientific Discoveries

Our knowledge of quantum mechanics is constantly expanding, providing fresh perspectives on the intricate connection between consciousness and reality. Once disregarded as a fringe field, quantum physics is now gaining recognition for its profound impact on our understanding and engagement with the world. Understanding ideas such as the observer effect and quantum consciousness uncovers the fascinating link between the mind and matter. Studies have revealed that our intentions have the power to shape real-world results, providing insights into how ideas transform into tangible outcomes.

As we delve into these concepts, it's important to base our understanding on scientific evidence. Various experiments have demonstrated the observer effect, where the mere act of observation has the power to influence the behaviour of particles. This observation reinforces the notion that our awareness has a substantial impact on shaping our encounters. Similarly, research on meditation and visualisation has revealed the profound impact these practices can have on the brain and body. This reinforces the idea that our thoughts and beliefs hold immense power in shaping our reality.

Scientific Studies Backing Quantum Manifestation

Various studies have corroborated concepts related to manifestation in quantum mechanics and psychology.

1. **Observer Effect in Quantum Experiments**: Multiple experiments, such as the double-slit experiment, demonstrate how observation might affect the study's results. Such experiments suggest that consciousness is the most fundamental constituent of reality, or at the very least, that our intentions and beliefs influence what occurs in our lives.

2. **The Placebo Effect**: Researchers have proven that belief and expectations play a crucial role in the placebo effect, leading

to genuine physical improvements felt by the patient. This demonstrates the power of the mind in influencing realities, showing that thoughts and emotions can indeed yield tangible results in life.

3. **Meditative/Visualisation Studies**: Scientific research has shown evidence that meditation and visualisation can ease mental and physical illnesses. These activities help avert stress and encourage wellness, thereby elevating the quality of life.

There is no necessary contradiction between the principles of quantum mechanics and those enunciated by Neville Goddard. What we need to do is synthesise scientific understanding with spiritual practice into one holistic approach to manifestation, capitalising on the strengths of both areas. This seems to affirm that reality is not some kind of ready-made, unchangeable construct but a dynamic field of possibilities materialised depending on the state of our consciousness. We can guide the collapse of the wave function towards our desired ends by aligning our thoughts, emotions, and beliefs with what we want in life, thus bringing our dreams into reality. It is remarkable how powerful imagination and intention can be.

Finally, 'Quantum Creation' aims to motivate and make one realise that the power of imagination and intention can deliver success. Reaching out to science and spirituality may offer a pathway towards deciphering the universe's codes to bring our desires to fruition. As we embark on this journey together, let us remain open to all possibilities and create a reality that captures the highest aspirations of our lives. We must remember that our understanding of the quantum world and its relation to the state of manifestation remains limited. Quantum mechanics is an elusive and shifting science, and the actual connection between the subatomic world and our classical reality is still an open question. In this section, therefore, we critique several considerations that critics have presented as evidence against the quantum premises of manifestation.

- **The Misconception of Quantum Phenomena**: One major criticism is that some proponents of quantum manifestation overstate or oversimplify accurate scientific phenomena. Features of the quantum domain, such as superposition and entanglement, do not translate straightforwardly into the everyday realm. It is crucial to distinguish well-established principles and empirical evidence in quantum mechanics from more speculative applications to manifestation practices.

- **The Lack of Replicable Experiments**: Unlike controlled experiments, which are the cornerstone of scientific inquiry, the results from manifestation practices are subjective and hard to replicate. External circumstances, belief systems, confirmation bias, or other factors can influence the outcomes of manifestation techniques. Thus, it is difficult to pin down cause and effect with exact actions and results.

- **The Peril of Oversimplification**: Quantum mechanics is a fascinating but complex science. Oversimplifying its principles into guidelines for manifestation can lead to misinterpretations and false expectations. Scientific ideas must be treated with depth to avoid such pitfalls. Nevertheless, the possible synergy of quantum mechanics with manifestation practices creates an exciting avenue for exploration.

Here are some salient points to guide this journey:

- **The Power of Possibility**: Quantum mechanics demonstrates that the universe is, to some extent, fluid and uncertain. This opens up the possibility for our thoughts, beliefs, and intentions to subtly yet significantly affect the trajectory of our lives.

- **Focus is Key**: The observer effect highlights the impact of focused attention. By holding our thoughts and emotions consciously on a desired endpoint, we can align our path towards its realisation.

- **Role of Belief**: Superposition suggests that an unwavering belief in a desired outcome may keep the energy open to many possibilities, thereby facilitating manifestation.

While these concepts are appealing, it is important to retain a degree of scepticism and a commitment to ongoing learning. Quantum mechanics is a rigorous science requiring continuous research and experimentation.

Manifestation practices represent an eclectic fabric of spiritual traditions and lived experiences. The goal here is to find common ground bridging the gap with integrative wisdom, providing a holistic perspective on personal growth and transformation.

The Purpose of This Book

This book strives to bring together two worlds that appear somewhat divergent: the mystic revelations of Neville Goddard and the new-wave theories of quantum mechanics. We hope that examining how these two realms meet will afford a deeper understanding of how we can consciously create our realities. Quantum physics is ensnaring in concept and counterintuitive in principle; its scientific framework provides an appropriate underpinning to complement the teachings Goddard sought to make people understand. Tenets like wave-particle duality, superposition, and entanglement provide interesting dimensions to view the entire manifestation process.

With those tenets in mind, one will see the power of thought, beliefs, and assumptions more clearly shaping our experiences' form and structure. The Law of Assumption by Neville Goddard offers something tangible through which one can tap into this power. His insistence on such concepts as imagination, self-concept, and the assumption's firmness or feeling of the wish fulfilled casts a blueprint upon which to base quantum applications in our everyday lives. Together, these insights create a powerful synergy that can radically transform our perception of how to interact with the world entirely.

This book is structured into two parts to provide a seamless journey from understanding foundational concepts to applying practical tools for quantum manifestation.

- **Part 1** introduces quantum creation's theoretical, philosophical, and scientific underpinnings. It ensures readers build a solid conceptual foundation by exploring the fundamentals of quantum theory, Neville Goddard's teachings, and the intersection of metaphysics and quantum mechanics. Chapters like "The Observer Effect" and "Superposition and Possibility" bridge complex quantum phenomena with practical insights into manifesting reality.

- **Part 2** focuses on practical tools, advanced techniques, and mastery of quantum manifestation. Actionable insights, such as visualisations, affirmations, and meditations, are explored alongside more advanced topics like neutrality, giving, and energy philanthropy. Chapters like "The Quantum Toolbox" and "Overcoming Common Blocks" bridge foundational knowledge and applied techniques.

As readers journey through this book, they may notice that specific themes—such as the Observer Effect, Superposition of Possibilities, Quantum Entanglement, and the Quantum Fabric of Creation—reappear in various chapters. These foundational concepts form the tapestry of quantum creation and are intricately woven throughout the narrative. However, beginning with Chapter 16, I have offered a more nuanced and focused exploration of various quantum perspectives, presenting them as standalone chapters for two reasons.

Firstly, this book is intended to be more than a linear narrative; it is designed to function as a menu of possibilities, much like dining at a fine restaurant. Each chapter is a dish crafted with care, allowing readers to select what most appeals to their current curiosity or need. These dedicated chapters offer a deep dive into their essence for those seeking an in-depth understanding of specific aspects—whether it be the Quantum Alchemy of Desire or the Symphony of Harmonics in tune with the universe.

Secondly, while these chapters build on earlier discussions, they are not repetitions. They are expansive explorations enriched with fresh insights and detailed treatments for those who wish to immerse themselves in particular facets of quantum creation. I have endeavoured to honour the diversity of my readers' interests, knowing that some may want to explore gratitude as a geometrical force of manifestation, while others may be drawn to the art of crafting realities through neutrality.

To ensure this journey remains engaging and avoids monotony, I've woven vivid descriptions of the Nilgiri Hills into the latter chapters, where much of the book was written. Nestled amidst mist-laden mountains, lush tea estates, and vibrant bursts of rhododendrons, the hills offered a serene backdrop that mirrored the principles of quantum creation—a landscape constantly recreating itself, just as we do through our thoughts and intentions. These moments of nature's grandeur are not mere digressions; they are reminders of how universal harmonics find balance, providing readers with a meditative pause amidst the depth of ideas. By grounding the abstract with the tangible, the sights and whispers of these verdant landscapes aim to rejuvenate the mind and uplift the soul, creating a more immersive and balanced reading experience.

Lastly, the decision to create separate chapters acknowledges the profound complexity and beauty of quantum creation itself. Just as light can be studied as both a particle and a wave, so too can these concepts be viewed through multiple lenses. Each chapter invites readers to approach these ideas anew, discovering layers of meaning that resonate with their unique journey.

In offering this rich tapestry, I hope to inspire understanding and wonder—a sense of discovery that mirrors the essence of quantum creation. Whether readers choose to savour a single chapter or consume the book as a whole, may they find within these pages the tools to craft their reality and the wisdom to see that every possibility is a doorway to infinite potential.

What to Expect From This Book

Below is a brief overview of what each chapter offers:

PART 1: FOUNDATIONS AND PRINCIPLES OF QUANTUM MANIFESTATION

1. **THE FOUNDATIONS OF QUANTUM THEORY**
   This chapter explains the basics of quantum mechanics in simple terms, covering essential concepts like wave-particle duality, superposition, and entanglement. It lays the groundwork for the rest of the book.

2. **THE POWER OF ASSUMPTION – UNLOCKING REALITY WITH NEVILLE GODDARD**
   Explore the transformative power of assumption, imagination, and the feeling of the wish fulfilled. Real-life success stories illustrate how these principles have changed lives.

3. **BRIDGING THE GAP: QUANTUM MECHANICS AND METAPHYSICS**
   This chapter connects quantum mechanics with metaphysical teachings, highlighting the observer effect, quantum consciousness, and their resonance with Goddard's philosophy.

4. **THE OBSERVER EFFECT: MANIFESTING REALITY**
    Learn how observation and intention influence quantum outcomes, with examples from experiments and practical applications in daily life.

5. **SUPERPOSITION AND POSSIBILITY: EXPANDING POTENTIAL**
    Discover how holding multiple possibilities in your mind can manifest desired outcomes. Real-life examples illustrate the power of embracing superposition.

6. **QUANTUM ENTANGLEMENT AND CONNECTEDNESS: MOVING BEYOND DUALITY IN MANIFESTATION**
    Explore the interconnectedness of thoughts, feelings, and distant outcomes through quantum entanglement, supported by scientific experiments and real-life stories.

7. **THE QUANTUM POWER OF INTENTION – SHAPING REALITY WITH THOUGHT**
    Understand the intricate web of energy and vibration that connects everything in the universe, revealing the essence of creation itself.

8. **THE QUANTUM EMBROIDERY: WEAVING NEVILLE GODDARD'S CONCEPTS INTO THE FABRIC OF QUANTUM PHYSICS**
    This chapter integrates Goddard's teachings with quantum physics, showing how the two fields complement each other to enhance manifestation practices.

9. **INTEGRATING SCIENCE AND SPIRITUALITY IN DAILY LIFE**
    Learn how to balance scientific understanding with spiritual practice, making quantum manifestation a natural part of everyday life.

15. **BECOMING NOBODY: A JOURNEY INTO THE ZERO-POINT FIELD**
Explore the zero-point field, where infinite possibilities exist, and learn how to tap into this state to manifest effortlessly.

16. **THE POWER OF SILENCE: UNLOCKING QUANTUM CREATION THROUGH STILLNESS**
Discover why silence is the key to unlocking infinite possibilities in the quantum field. Learn how stillness nurtures the seeds of desire, aligning your inner state with the creative forces of the universe.

17. **ENERGY MANAGEMENT: HARNESSING QUANTUM POWER IN THE BLUE MOUNTAINS**
Dive into the profound connection between energy, consciousness, and reality creation. Discover how energy flows within and around you and how you can harness it to shape your life deliberately.

18. **THE QUANTUM SYMPHONY: TUNING INTO THE UNIVERSE'S HARMONICS**
Align with the universe's natural frequencies to amplify your manifestation power.

19. **THE LAW OF TRINITY: GRATITUDE, RESET, AND THE GEOMETRY OF MANIFESTATION**
Explore the transformative power of gratitude and reset your intentions with insights into the geometry of manifestation.

20. **THE ALCHEMY OF CONTENT CREATION: MANIFESTING INFLUENCE AND AUTHENTICITY**
Authenticity in your creative endeavours using quantum principles.

CONCLUSION: THE FUTURE OF QUANTUM MANIFESTATION

The book concludes by tying together all the concepts, offering a vision for the future of quantum manifestation and practical guidance for continued growth.

PART 1

Foundations and Principles of Quantum Manifestation

CHAPTER 1

The Foundations of Quantum Theory

Introduction: The Dawn of Quantum Creation

It was a clear, crisp evening in the late 1940s, and a young woman stood by the windowsill of her modest New York apartment, gazing out at the city lights flickering like distant stars. She had just finished reading a book by an obscure but intriguing figure, Neville Goddard, who spoke of the immense power of the imagination and the transformative nature of assuming one's desires as already fulfilled. The words resonated with her deeply, and though the world outside her window seemed indifferent, she tested Goddard's theory.

Every night before sleep, she would close her eyes and imagine herself in the life she desired—not in the future, but as if it were already happening. She felt that life's emotions lived in their fullness within her mind, and then she let them go.

A few months later, her life changed in ways she had never expected. Opportunities seemed to appear out of nowhere, and doors that had once been closed opened. Her imagined life, which remained confined to her mind, began manifesting in her reality.

The young woman's story became one of the many testimonials in Goddard's teachings. Still, it also sparked an interest that would echo through the decades: What if a scientific explanation existed behind this seemingly mystical process? Could modern physics, particularly the strange and often counterintuitive principles of quantum mechanics, provide a framework for understanding how our thoughts and intentions shape our reality?

The Foundations of Quantum Theory

In the small town of Ulm, Germany, on a warm March day in 1879, a child who would one day change the course of human understanding was born. His name was Albert Einstein, and his theories of relativity would revolutionise how we perceive space, time, and energy. Yet, Einstein's work would also pave the way for the development of quantum mechanics—a branch of physics that explores the strange and often bewildering behaviour of particles at the smallest scales of existence.

But before we delve into the complexities of quantum mechanics, let us take a step back and consider the world that existed before Einstein's birth—a world governed by the classical mechanics of Isaac Newton. For over two centuries, Newton's laws provided a solid framework for understanding the motion of objects, the forces that govern them, and the predictability of cause and effect. According to classical physics, the universe was a vast, well-oiled machine where every action had an equal and opposite reaction. We could determine the future with precision given the correct information.

However, as scientists peered deeper into matter at the turn of the 20th century, cracks appeared in the Newtonian worldview. Experiments revealed behaviours that defied classical explanations, and the quest to understand these anomalies gave birth to the wonderful world of quantum mechanics.

The Birth of Quantum Mechanics: Max Planck's Quanta

Our journey into quantum theory began in 1900 with a German physicist named Max Planck. Picture an old, wise man sitting by a fire, pondering the mysteries of the universe. Planck was trying to figure out why objects, when heated, emit light in different colours—like how a piece of metal glows red, then yellow, and eventually white as it gets hotter.

Classical physics couldn't explain this, so Planck proposed something radical: What if light, instead of being a continuous wave, comprised tiny, discrete packets of energy? He called these packets 'quanta'. This idea was like planting the first seed in what would grow into the vast tree of quantum mechanics.

Planck's discovery set the stage for other physicists to explore the strange behaviour of these quanta, leading to a revolution in how we understand the universe at its smallest scales.

Wave-Particle Duality: The Dual Nature of Reality

One of the earliest challenges to classical physics came from the study of light. For centuries, scientists debated whether light comprised particles or waves. Newton believed light comprised tiny particles, or 'corpuscles'. In contrast, others, like the Dutch physicist Christiaan Huygens, argued that light behaved like a wave, spreading out and interfering with itself.

The debate seemed to be settled in the 19th century when Scottish physicist James Clerk Maxwell showed that light was an electromagnetic wave capable of travelling through the vacuum of space. However, this wave theory became questionable due to experiments conducted in the early 20th century, most notably the photoelectric effect—a phenomenon in which light shining on a metal surface could eject electrons from that surface.

The problem was that the wave theory of light could not fully explain the results. Einstein proposed that while behaving as a wave in some contexts, light also comprised discrete packets of energy called 'quanta' or 'photons'. These photons carried energy that could knock electrons loose from the metal surface, depending on their frequency—a concept that would later earn Einstein the Nobel Prize in Physics.

This dual nature of light, known as wave-particle duality, was one of the first indications that the universe at the quantum level did not conform to the neat categories of classical physics. Light could behave as a wave, spreading out and interfering, and as a particle, delivering energy in discrete packets. This discovery would lay the groundwork for further explorations into the dual nature of reality itself.

The Double-Slit Experiment: A Window into Quantum Mysteries

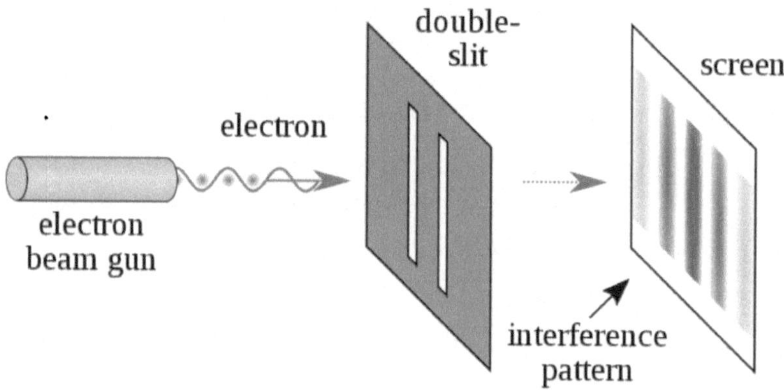

If wave-particle duality was a crack in the classical worldview, the double-slit experiment was a full-blown fissure. First conducted by Thomas Young in 1801 and later revisited in quantum mechanics, the double-slit experiment is perhaps one of the most famous and puzzling experiments in the history of science.

The experiment involves shining a beam of light through two parallel slits and observing the pattern that forms on a screen behind them. When both slits are open, the light behaves as a wave, creating an interference pattern of alternating bright and dark bands on the screen—evidence that the waves from each slit are overlapping and interfering with each other.

However, something extraordinary happens when the experiment involves particles like electrons. If the particles get fired through the slits one at a time, one might expect them to behave like tiny bullets, passing through either slit and forming two distinct bands on the screen. But this is not what occurs. Instead, even when particles are sent through the slits one at a time, they still produce an interference pattern, as if each particle is passing through both slits simultaneously and interfering with itself.

To make matters even stranger, when scientists place detectors at the slits to observe which slit the particle passes through, the interference pattern disappears. The particles behave as expected, forming two distinct bands. This suggests that mere observation collapses the wave-like behaviour into particle-like behaviour—a phenomenon later known as the observer effect.

The double-slit experiment challenges our most fundamental assumptions about reality. It suggests that particles exist in a superposition state, occupying multiple states simultaneously until observed. It also implies that observation plays a crucial role in determining the outcome of an experiment, blurring the line between observer and observed.

Superposition: The Quantum Realm of Possibilities

The concept of superposition is a cornerstone of quantum mechanics and a key to understanding the vast potential inherent in the universe. In classical physics, an object is in one state or another—a coin is heads or tails, and a cat is alive or dead. But in the quantum realm, particles exist in a state of superposition, where they can be in multiple states simultaneously.

The most famous illustration of superposition comes from the thought experiment proposed by Austrian physicist Erwin Schrödinger, known as Schrödinger's cat. In this scenario, a cat is placed in a sealed box with a vial of poison, which is released if a radioactive atom decays. According to quantum mechanics, the atom exists in a superposition of decayed and undecayed states until observed, meaning the cat is simultaneously alive and dead until someone opens the box.

While Schrödinger's cat was intended to critique the Copenhagen interpretation of quantum mechanics, it has become a powerful metaphor for the strange and counterintuitive nature of quantum superposition. In the quantum world, particles do not exist in a single state but in a superposition of all possible states, and only when they are observed do they collapse into a definite state.

This concept has profound implications for our understanding of reality and manifestation potential. If particles can exist in multiple states until observed, then reality is not fixed but fluid, shaped by observation. This idea resonates deeply with Neville Goddard's teachings on the power of imagination and the role of consciousness in shaping reality.

Stern-Gerlach Experiment: The Birth of Quantum Spin

Now, let's pause our journey to visit an old laboratory in Germany in 1922, where two scientists, Otto Stern and Walther Gerlach, were conducting an experiment that would add another piece to the quantum puzzle. They were curious about how particles behaved when they passed through a magnetic field. So, they fired silver atoms through a magnetic field to see what happened.

In classical physics, you might expect the particles to move in a straight line or scatter in predictable ways. But what they discovered was surprising: the particles split into two distinct paths, which showed that the particles had a property called 'spin', which could be up or down but not both simultaneously.

Imagine you're at a carnival and come across a game where you have to shoot metal balls through a magnetic field to win a prize. Instead of rolling through as you'd expect, the balls start acting strangely. Some go up, and some go down when they pass through the magnetic field, even though you rolled them the same way!

It's surprising because, in our everyday world, things rarely behave like this. Stern and Gerlach discovered that these tiny particles have something like a little magnet inside them, and this 'magnet' can only point up or down—never in between. This experiment showed that particles can behave in bizarre and unexpected ways in the tiny world of quantum mechanics, which doesn't align with classical expectations. It added another layer to our understanding of the quantum world, where particles behave in fascinating and puzzling ways.

Entanglement: The Web of Interconnectedness

Another cornerstone of quantum mechanics is the phenomenon of entanglement, which Einstein famously called 'spooky action at a distance'. Entanglement occurs when two particles become linked so that the state of one particle becomes instantly correlated with the state of the other, regardless of the distance between them. So, if you measure the state of one entangled particle, you immediately know the state of its partner, even if it is on the other side of the universe.

Entanglement challenges our classical notions of locality and causality, suggesting that the universe is far more interconnected than we once thought. It also opens the door to the possibility that our thoughts and intentions, which are forms of energy, can influence distant outcomes in ways that defy classical explanations.

One of the most famous discussions of entanglement came from the EPR paradox, named after Einstein, Podolsky, and Rosen. These three scientists proposed a thought experiment in 1935 to highlight what they saw as a problem with quantum mechanics. They argued that if quantum mechanics were correct, it would mean that two particles could instantaneously affect each other, even across vast distances, which seemed impossible.

Imagine now, let's say you have two identical toys. If you twist one, the other does the same thing, even if they are on opposite sides of the playground! That would be pretty amazing, right?

However, in the 1960s, physicist John Bell developed Bell's theorem, which tested whether this 'spooky action' was real. His work led to experiments, such as those conducted by Alain Aspect in the 1980s, confirming that entanglement was genuine. These experiments showed particles could connect in ways that defy classical explanations and that reality at the quantum level is deeply interconnected.

Imagine you and your friend have two magical dice. If you roll them, no matter how far apart you are, the dice always show the same number. You might think there's some trick involved, maybe a secret connection between the dice.

John Bell devised a way to test whether these dice worked by some hidden trick or were just magically connected. His idea was called 'Bell's theorem'. Bell's theorem showed that if the particles secretly communicated through some hidden information, they wouldn't behave as quantum mechanics predicted.

Fast forward to the 1980s. A scientist named Alain Aspect tested Bell's ideas with experiments. He used pairs of tiny light particles called photons. He separated them and measured their properties. Like those magical dice, when Aspect measured one photon, the other one always seemed to know what was happening instantly, even when they were far apart.

Aspect's experiments showed the particles got connected in a way that didn't involve any hidden tricks. They were just 'entangled', meaning their states were mysteriously linked, no matter how far apart. This experiment helped prove that quantum mechanics was correct and that the universe at the tiniest levels behaves in bizarre and fascinating ways.

The EPR paradox, Bell's theorem, and Alain Aspect's experiments all showed that once particles get entangled, they remain connected, and a change in one will instantaneously affect the other, regardless of distance, suggesting that everything in the universe connects in a way that transcends classical physical boundaries. This interconnectedness has led to the idea that our thoughts, actions, and observations might influence distant events or outcomes, highlighting the profound unity and coherence within the fabric of reality.

The implications of entanglement are profound. They suggest the universe is not a collection of separate, isolated parts, but a web of interconnected relationships. This interconnectedness is not just a physical phenomenon; it also resonates with spiritual teachings about the oneness of all things.

Together, these principles establish a vision of the universe as a dynamic, interconnected web of possibilities, where the boundaries between objects and states are not as clear-cut as they appear in classical physics. These insights revolutionise our understanding of the physical world and open the door to new ways of thinking about reality, consciousness, and the power of the mind.

Real-Life Applications: Quantum Biology, Cryptography, and Computing

Quantum mechanics may seem like an abstract field, but its principles have real-world applications that are transforming our lives. For example, in quantum biology, researchers are exploring how quantum effects like superposition and entanglement play a role in photosynthesis, where plants convert sunlight into energy with remarkable efficiency. These discoveries could lead to new technologies that mimic these processes to create more efficient solar cells.

Quantum cryptography uses quantum mechanics principles to create secure communication systems. Because observing a quantum system alters its state, it is possible to generate encryption methods that are theoretically unbreakable, ensuring the secure transmission of information.

Quantum computing is perhaps one of the most exciting applications of quantum mechanics. Unlike classical computers, which process information in binary bits (0s and 1s), quantum computers use quantum bits, or qubits, which can exist in multiple states simultaneously because of superposition. This allows quantum computers to perform complex calculations at speeds far beyond the capabilities of classical computers, potentially revolutionising fields such as medicine, finance, and artificial intelligence.

One famous example of quantum computing's potential is Google's achievement of 'quantum supremacy' in 2019, where their quantum computer performed a task in 200 seconds that would have taken the world's fastest supercomputer 10,000 years to complete. This milestone marked a significant step forward in harnessing the power of quantum mechanics for practical applications.

Conclusion: The Bridge Between Science and Spirituality

The journey we have embarked upon in this chapter has taken us from the early days of classical physics to the strange and wondrous world of quantum mechanics. We have explored the fundamental concepts of wave-particle duality, superposition, entanglement, and the observer effect—concepts that challenge our classical understanding of reality and open the door to new possibilities.

These principles of quantum mechanics reveal a far more mysterious and interconnected universe than we once imagined. They suggest that reality is not a static, predetermined machine but a dynamic, fluid field of possibilities shaped by observation, intention, and interconnectedness.

References

1. Max Planck, *The Origin and Development of Quantum Theory*, Clarendon Press, 1925.
2. Albert Einstein, *Relativity: The Special and the General Theory*, Methuen & Co. Ltd., 1916.
3. Erwin Schrödinger, *What is Life? The Physical Aspect of the Living Cell*, Cambridge University Press, 1944.
4. Neville Goddard, *The Power of Awareness*, Prentice-Hall, 1952.

Articles and Papers:
1. John S. Bell, "On the Einstein Podolsky Rosen Paradox," *Physics Physique Физика*, vol. 1, no. 3, 1964, pp. 195-200.
2. Alain Aspect, Philippe Grangier, and Gérard Roger, "Experimental Realisation of Einstein-Podolsky-Rosen-Bohm Gedankenexperiment: A New Violation of Bell's Inequalities," *Physical Review Letters*, vol. 49, no. 2, 1982, pp. 91-94.
3. Albert Einstein, Boris Podolsky, and Nathan Rosen, "Can Quantum-Mechanical Description of Physical Reality Be Considered Complete?" *Physical Review*, vol. 47, 1935, pp. 777-780.

CHAPTER 2

The Power of Assumption – Unlocking Reality with Neville Goddard

Remember, the world reflects your own desires. Ask yourself what you want, and then make it happen. Don't worry about the details of how it will happen; just trust that the results will manifest themselves in due time.

– Neville Goddard

Introduction

Neville Goddard, a figure of great intrigue and profound wisdom, has left an indelible mark in the realms of spiritual philosophy and self-development. Born in 1905 in Barbados and passing in 1972, Neville's teachings have left a lasting impact on people worldwide, transcending time and space. Neville was an unconventional teacher, someone who dared to explore the depths of the human mind and look beyond the boundaries of the material world.

Neville's teachings centre on a fundamental concept: The Law of Assumption. This principle suggests that the world we perceive is a mirror of our inner thoughts and feelings. It implies that our assumptions, beliefs, and imagination shape our reality. Neville put forth the idea that by embracing the sensation of our desires already being fulfilled, we have the power to manifest them in the physical world. It's a thought-provoking idea that defies traditional reasoning and encourages us to explore the deeper, philosophical aspects of life.

The Law of Assumption: Manifesting Reality from Within

At the core of Neville's philosophy lies the Law of Assumption, a principle that proposes our reality is not a rigid construct but a fluid, malleable reflection of our internal world. According to Neville, our assumptions play a significant role in shaping our external experiences. If we embrace a positive mindset, our external reality will reflect that positivity. If we embrace empowering beliefs and positive assumptions, our reality will mirror those possibilities.

Let's delve into the fascinating tale of Jim Carrey, a renowned actor, comedian, and filmmaker, to illustrate the incredible influence of assumption. Prior to achieving widespread recognition, Carrey was an aspiring actor facing the challenges of breaking into the competitive world of Hollywood. However, he refused to let his present circumstances determine his future. Instead, Carrey wrote himself a cheque for $10 million, dated five years into the future, and kept it in his wallet. He would imagine receiving that amount for his acting services, fully immersing himself in the sensation of already being an accomplished actor. By a twist of fate, right before the date on the cheque, Carrey scored a role in *Dumb and Dumber* that earned him a cool $10 million.

Carrey's story showcases the strength of assumption. With unwavering faith in his own abilities, he harmonised his inner world with his aspirations, ultimately manifesting them in the physical realm. According to Neville, the world reflects our internal beliefs and assumptions.

Imagination: The Workshop of the Mind

Neville Goddard frequently described imagination as the 'workshop of the mind', where one brings all things into existence. He strongly believed that imagination is not merely a tool for creativity but the very essence of reality. According to Neville, the power of our imagination and firm belief can bring about the manifestation of our desires.

One of the most fascinating stories that exemplifies this principle is that of Nikola Tesla, a genius inventor, electrical engineer, and physicist. Many people recognise Tesla for his contributions to modern electrical engineering, but they often overlook the role of his vivid imagination in bringing his inventions to life. Tesla confidently asserted that he possessed the extraordinary ability to envision his inventions vividly and with remarkable precision, eliminating the need for tangible prototypes. He would meticulously craft and perfect these inventions solely within the realm of his imagination before bringing them to life in the tangible world.

Tesla's approach is in perfect harmony with Neville's teachings on the power of imagination. Tesla relied not only on imagination for his inventions; his vivid imagination enabled him to create some of the most revolutionary technologies of our time.

Understanding Self-Concept: Building Your Reality

Neville highlighted the significance of self-concept—how we view ourselves—as a crucial element in the process of bringing things into reality. He believed that our self-concept has a profound impact on various aspects of our lives, including our relationships, careers, health, and financial well-being. If we have a positive self-perception and believe in our own worth and capabilities, the world will respond in kind, shaping our everyday lives.

An intriguing illustration of this principle in action is the tale of Mahatma Gandhi, the revered figure who spearheaded the Indian independence movement against British rule. Gandhi believed in the power of nonviolent resistance to bring about significant change. Against all odds, Gandhi's unwavering faith in his mission and his own leadership abilities shaped the course of history by leading India to independence. His firm belief in peaceful methods of resistance was clear in the nonviolent movement that played a crucial role in India gaining independence in 1947.

Gandhi's life exemplifies how our self-concept can influence the world, showing that it is not merely a private, internal matter. By nurturing a confident and optimistic self-image, we have the power to shape not just our own lives but also the world.

Embracing the Fulfilment of Desires

One technique that Neville taught is the practice of assuming the feeling of the wish fulfilled, which can be powerful. It goes beyond simply imagining what you want; it's about fully immersing yourself in the belief that your desire has already come true, feeling the emotions and sensations that come with its fulfilment.

Let's delve into the fascinating tale of Andrew Carnegie, the Scottish-American industrialist who rose to become one of the most affluent individuals in history. Many often credit Carnegie's remarkable success to his sharp business sense and strategic mindset. However, people often overlook his unshakeable confidence in his own potential for greatness. Carnegie embraced a mindset of success well before it became a reality, following Neville's Law of Assumption. He would dedicate moments every day to envisioning his triumphs, embracing the emotions of achievement and abundance, and carrying that sense of fulfilment with him all day long.

By embracing the power of envisioning his desires as already accomplished, Carnegie could conquer countless hurdles and difficulties on his journey to becoming a formidable force in the business world. Through the power of his mindset, Carnegie created a life filled with remarkable achievements and accomplishments.

The Power of Belief: A Universal Principle

At the core of Neville Goddard's teachings lies a profound principle: the power of belief in shaping our reality. Our beliefs shape the experiences we attract into our lives, whether or not we realise it. Neville made a compelling case that mere wishing is insufficient; we must wholeheartedly believe that our desires are already within our grasp.

Helen Keller, the remarkable American author, disability rights advocate, and lecturer, serves as a powerful testament to the incredible strength of belief. Despite being deafblind, she defied all odds and became the first person in her situation to earn a Bachelor of Arts degree. Even though she lost her sight and hearing at a young age because of illness, Keller never lost her belief in her own potential to learn, communicate, and make a difference in the world. Anne Sullivan, Keller's teacher, played a crucial role in helping her learn to read, write, and speak. Thanks to Sullivan's guidance, Keller became a highly influential author and advocate.

Keller's life is a shining example of how belief can truly transform one's existence. Her firm belief in her own abilities enabled her to conquer seemingly impossible obstacles and leave a lasting mark on the world. Keller's story beautifully reflects Neville's teachings on the transformative power of belief. It serves as a powerful reminder that with unwavering faith in our own abilities, we can conquer any obstacle that comes our way.

The Importance of Faith and Perseverance

Neville emphasised the importance of having faith and staying persistent in order to bring about the desired outcomes. Embracing the sensation of the desired outcome consistently is crucial—not just occasionally, but persistently. We must hold on to our belief, even when faced with challenges.

One of the most captivating stories of unwavering determination and belief is the tale of Walt Disney. Prior to establishing the worldwide entertainment empire that now bears his name, Disney encountered a series of setbacks and failures. He faced obstacles in his career, including being let go from a newspaper job for 'lacking imagination' and experiencing the bankruptcy of his first animation company. Despite these setbacks, Disney remained steadfast in his conviction that he could bring forth something truly remarkable. He imagined his aspirations, held onto his beliefs, and persevered in his efforts despite the challenges he faced.

Disney's unwavering dedication and determination eventually paid off, resulting in the creation of cherished characters, films, and theme parks that have brought immense joy to millions across the globe. His story perfectly illustrates Neville's teachings on the power of persistence and assumption in manifesting one's desires.

Real-Life Applications: Success Stories
Conclusion: Embracing the Power Within

Neville Goddard's teachings convey a powerful and uplifting message: we possess the ability to mould our reality by harnessing the creative potential of our imagination, nurturing a positive self-image, and embracing the sensation of our desires already being fulfilled. These principles encompass more than just abstract ideas.

However, these principles apply to everyone, not just the famous and successful. They are accessible to all, regardless of background or circumstances. It's important to grasp the concept that reality is not a rigid, unchanging structure but a mirror of our inner world. Through a shift in our assumptions, beliefs, and self-concept, we have the power to transform our reality.

As you delve deeper into this book, I invite you to explore these principles in your own life. Begin by examining your assumptions and beliefs. Do they align with the reality you desire to create? If not, start intentionally changing them. Let your imagination run wild as you paint a picture of your desired outcomes. Embrace the feeling of your wishes coming true and practice embodying that sensation.

Always remember that you can shape your own reality. By embracing Neville Goddard's teachings, you can tap into the incredible power of your mind and bring forth a life that mirrors your greatest dreams and ambitions.

References

1. Goddard, Neville. *The Power of Awareness*. Martino Fine Books, 2013.
2. Tesla, Nikola. *My Inventions: The Autobiography of Nikola Tesla*. BN Publishing, 2006.
3. Seifer, Marc J. *Wizard: The Life and Times of Nikola Tesla*. Citadel, 2001.
4. Keller, Helen. *The Story of My Life*. Bantam, 1990.
5. Disney, Walt. *The Art of Walt Disney: From Mickey Mouse to the Magic Kingdom*. Harry N. Abrams, 2006.

CHAPTER 3

Bridging the Gap: Quantum Mechanics and Metaphysics

Introduction

It was a quiet morning in Chennai, where the city seemed to hold its breath before the hum of life crescendoed. From the veranda, I watched the first light of dawn spill over Marina Beach, painting the waves in hues of gold and silver. The rustling of coconut palms mixed with the faint calls of chai vendors setting up their stalls along the coast. In the moment's stillness, my thoughts drifted to the extraordinary dance of the universe—a dance governed by principles as ancient as the cosmos and as groundbreaking as quantum mechanics.

This chapter embarks on a journey to uncover the connection between the enigmatic laws of the quantum world and the timeless metaphysical insights of Neville Goddard. Much like the city of Chennai, where modernity blends seamlessly with tradition, these two seemingly disparate realms converge, offering profound lessons on the nature of reality. We will explore phenomena such as the observer effect, wave-particle duality, and quantum entanglement, not as abstract theories confined to laboratories but as doorways to understanding our most profound potential. Together, we'll bridge the chasm between science and spirituality, weaving their insights into a tapestry of practical wisdom that enriches our daily lives.

Quantum Mechanics: The Science of the Very Small

Quantum mechanics, a fascinating field of physics, explores the intricate behaviour of particles at the atomic and subatomic levels. This field has completely disrupted traditional ideas in physics, unveiling a world that is incredibly bizarre and intricate beyond our wildest imagination. The principles of quantum mechanics push the boundaries of our conventional understanding of reality, introducing concepts that are both captivating and puzzling.

The Observer Effect: The Power of Perception

The observer effect is a fascinating concept that arises from the intricate world of quantum mechanics. It implies that simply observing a quantum system can change its state, causing it to transition from a range of possibilities to a single, definite result. This concept has far-reaching consequences, not only in the realm of science but also in our perception of the very nature of existence.

Neville Goddard's teachings on the power of assumption align closely with the observer effect, creating a profound resonance. According to Goddard, our perceptions, beliefs, and assumptions play a significant role in shaping our reality. The observer effect reflects the concept that the act of observation determines the outcome of a quantum experiment.

Sarah Blakely, the visionary behind Spanx, provides a fascinating real-life illustration of the observer effect in action. Before Blakely achieved her remarkable success, she started her career by selling fax machines door-to-door. Regardless of her situation, she firmly believed in her destiny for greatness. She imagined herself as a thriving entrepreneur and sensed the actuality of that triumph even before it materialised. Her strong conviction and determination toward her vision seemed to have a magical effect, turning countless possibilities into the tangible reality of her success.

Blakely's story is a compelling illustration of how the observer effect and the Law of Assumption can collaborate to mould reality.

Her success was not only attributed to her diligent efforts and inventive thinking but also to her unwavering self-confidence in her abilities. By embracing the power of positive thinking, she could shape her life in extraordinary ways, tapping into forces of the universe that go beyond our understanding.

Exploring the Intersection of Quantum Mechanics and Metaphysics

The principles of quantum mechanics offer a captivating scientific framework that resonates with many metaphysical concepts. For example, the concept of superposition resonates with the metaphysical idea of infinite possibilities. In the fascinating realm of quantum physics, particles exist in a realm of endless possibilities, where multiple outcomes coexist until an observation brings forth a singular reality. In metaphysics, believers maintain that the future is not fixed but encompasses a vast array of possibilities. By directing our attention and intention towards a desired outcome, we have the power to manifest that possibility into reality.

Quantum entanglement, too, shares intriguing connections with the metaphysical realm. The phenomenon of entangled particles reveals the profound interconnectedness that exists throughout the universe. This idea demonstrates a deep understanding of the interconnectedness of everything in the universe. It suggests that our thoughts, emotions, and actions have a profound impact that extends beyond ourselves.

The Intersection of Quantum Mechanics and Metaphysics

The intersection of quantum mechanics and metaphysics is clear in the fascinating realm of quantum consciousness. Some intriguing theories suggest that consciousness might be a quantum phenomenon. This concept proposes that our minds function on a quantum level, where thoughts and consciousness have the potential to shape the physical realm. If this is indeed the case, it suggests that the mind plays a vital role in shaping our perception of reality, actively contributing to its creation.

Practical Applications: Manifesting Through Quantum Metaphysics

Recognising the link between quantum mechanics and metaphysics goes beyond mere intellectual exploration; it holds practical significance for our daily existence. Considering the potential impact of our thoughts and observations on reality, it is crucial for us to be aware of our mental and emotional states.

Visualisation is a practical application that can be incredibly useful in many situations. It is a powerful tool commonly employed in various fields such as sports, business, and personal growth to create a vivid mental picture of the desired outcome. Through the power of imagination and the ability to feel the emotions tied to our desired outcome, we become the observers of a quantum experiment, influencing the manifestation of our reality.

Take Elon Musk, the mastermind behind companies like Tesla and SpaceX, who has emphasised the significance of dreaming big and envisioning triumph. Musk's visionary mindset has propelled him to achieve remarkable feats that were once considered impossible, such as envisioning a future dominated by electric cars and a humanity that spans multiple planets. His achievements showcase the practical application of quantum mechanics in transforming abstract concepts into tangible creations.

Another effective technique to consider is the use of affirmations. Affirmations are powerful tools that can help us shape our beliefs and align our subconscious minds with our desires. Through the regular practice of affirmations, we have the power to reshape our mindset and direct our attention towards the positive outcomes we desire, ultimately shaping the reality we encounter.

Appreciation is a potent tool in the realm of quantum metaphysics. By appreciating the good things in our lives, we open ourselves up to positive energy and invite more positive experiences into our lives. This concept is reminiscent of resonance in quantum mechanics, where particles can resonate with each other at specific frequencies.

Meditation is an essential practice in quantum metaphysics. By practising meditation, we can achieve a state of tranquillity, tap into the limitless potential of the universe, and channel our intentions towards our goals. Meditation opens up a world of endless possibilities, helping us connect with the quantum field and shape our desired reality.

The principles of quantum metaphysics are not just theoretical—they have been put into practice by numerous successful individuals to bring their desires to life. Here are some genuine examples:

Arnold Schwarzenegger: Prior to becoming a worldwide sensation, Schwarzenegger envisioned his triumphs in the realms of bodybuilding and acting. He frequently imagined himself clutching the Mr. Olympia trophy well before he even entered the competition, and he saw himself as a renowned Hollywood star long before he landed his major opportunity. Schwarzenegger's mastery of visualisation and the Law of Assumption enabled him to turn his dreams into tangible achievements.

Lewis Hamilton: In his career, Formula 1 racing champion Lewis Hamilton has emphasised the significance of mindset and visualisation. Prior to races, Hamilton would mentally envision the track, his driving strategy, and ultimately standing triumphantly on the podium as the victor. His exceptional mental preparation has played a crucial role in his remarkable success in Formula 1.

Final Thoughts: A Fresh Perspective

As we explore the fascinating connection between quantum mechanics and metaphysics, we discover a new paradigm of understanding. In this paradigm, science and spirituality are not conflicting but harmonious forces that can collaborate to enhance our lives. The wonders of the quantum world allow us to glimpse the profound mechanisms of the universe, where the mind and matter are intricately connected and where our thoughts and intentions can shape our reality.

The blending of these areas in my life has been a constant source of inspiration and empowerment. Throughout my journey to becoming a police officer, my experiences as an author, and my adventures as a marathon runner, I have had the privilege of witnessing the incredible influence of the mind in shaping our reality. Quantum mechanics has not only enhanced my understanding of the world but has also given me a practical framework to apply metaphysical teachings in transformative ways.

As you delve into the pages of this book, I encourage you to approach these ideas with an open mind and a genuine sense of curiosity. In the upcoming chapters, we will explore the fascinating realms of quantum mechanics and metaphysics, providing you with practical techniques to tap into their incredible potential and apply them to your everyday life. Let's embark on a fascinating exploration of the quantum field, where we'll uncover the mysteries of the universe and discover endless possibilities.

As we progress, let's remember the wise words of physicist Niels Bohr, who once said, *"If quantum theory doesn't surprise you, you haven't understood it."*

With this newfound perspective, life becomes a captivating journey filled with endless possibilities. We can embrace the power within us to shape the world and leave a lasting impact.

CHAPTER 4

The Observer Effect: Manifesting Reality

Observing an outcome has the power to alter the outcome itself. Imagine the incredible potential of the human mind: the ability to mould reality with a single thought. It's truly awe-inspiring.

The Art of Observation: Shaping the World Around Us

In the vibrant city of Chennai, I frequently strolled along Marina Beach, mesmerised by the soothing rhythm of the waves caressing the shore. The boundless sea never failed to evoke a sense of wonder, reminding me of the limitless potential that life offers. It was during one of these introspective walks that I discovered a profound link between quantum physics and the principles of manifestation—a connection that would ultimately inspire the birth of this book.

As I stood by the shore, mesmerised by the rhythmic dance of the waves, I couldn't help but ponder the interconnectedness of our existence. I realised our mere presence on the beach has a subtle but profound impact on the ebb and flow of the sand and water. In a similar vein, I couldn't help but reflect on how our thoughts and intentions shape the very fabric of reality, even if we may not always be aware of it. It's a fascinating and awe-inspiring realisation.

This realisation reminded me of my early days of delving into the realm of quantum mechanics. I recalled my fascination with the double-slit experiment and how it captivated me to think that the mere act of observing a particle could influence its behaviour. It went beyond mere scientific curiosity; it provided a deep understanding of the very essence of reality.

The Observer Effect, also known as the act of observation, has a fascinating impact on the outcome of events. It suggests that our perceptions and intentions shape the reality we experience. This concept resonates with the teachings of Neville Goddard, who believed that our consciousness has the power to create our reality. The Observer Effect provides a scientific foundation for Goddard's ideas, making them even more intriguing.

The Observer Effect in Daily Life: The Broader Implications

The Observer Effect extends beyond the boundaries of the quantum realm and holds significant implications for our daily lives. If our thoughts and observations can impact the behaviour of particles, they can also shape the broader reality we encounter. This concept serves as a connection between quantum mechanics and metaphysics, proposing that we are not mere spectators of our reality but engaged contributors in its formation.

Sachin Tendulkar, the iconic Indian cricketer, illustrates this concept. His remarkable concentration and unshakeable confidence in his own abilities marked Tendulkar's career. Prior to taking the field, he would vividly imagine his innings, envisioning the ball approaching him, the impeccable timing of his shot, and the resounding applause of the spectators as the ball sailed to the boundary. This mental practice served as more than just a method of preparation; it was a means for Tendulkar to shape the reality he desired.

Tendulkar's career is a shining example of the incredible impact of keen observation. His unwavering ability to perform at the highest level, even in the face of intense pressure, is credited to his deep understanding of the Observer Effect. By visualising success with unwavering clarity, he could bring that success to life on the field.

The Observer Effect and Manifestation: A Scientific Basis for Spiritual Teachings

The relationship between the Observer Effect and the teachings of Neville Goddard presents a fascinating narrative that harmonises science and spirituality. Goddard's core belief, the Law of Assumption, suggests that by embracing the sensation of our desires already being fulfilled, we can manifest them in our lives. This concept closely resonates with the principles of quantum mechanics, where the act of observation influences the outcome of a quantum event.

Have you ever wondered how these abstract principles actually work in real life? It all comes down to how we direct our thoughts and intentions. Just like how the observer in the double-slit experiment influences the outcome, our focused attention has the power to shape the reality we want.

Visualising our desired outcome with unwavering focus is crucial. By vividly imagining and holding that image in our minds, we can effectively bring it into existence. This practice goes beyond dreaming; it is a deliberate and intentional act of creation.

Let's look at the story of **A.R. Rahman,** the incredibly talented Indian composer and music producer. Despite facing many challenges and setbacks along the way, Rahman never lost sight of his goal. Late nights spent in his studio, enveloped in his music, Rahman envisioned his creations being embraced and loved by people all over the world. This unwavering determination, coupled with his undeniable talent, ultimately propelled him to the top of the music industry.

Rahman's story is a compelling illustration of the Observer Effect at work in the realm of manifestation. Through his unwavering vision of success, he manifested his dreams, triumphing over challenges and surpassing expectations.

Practical Techniques for Harnessing the Observer Effect

Grasping the Observer Effect is just the beginning; putting it into practice in our daily lives is a whole other ballgame. To truly tap into the potential of this phenomenon, we need to master the art of directing our thoughts and intentions with purpose and consistency. Here are a few handy techniques to help you do just that:

1. **Visualisation:** As I mentioned earlier, visualisation is a powerful tool for manifestation. Take a few moments each day to immerse yourself in the vivid details of your desired outcome. Picture every aspect of the experience—what you see, hear, feel, and even smell. By including as many sensory details as possible, your visualisation will become more tangible and ultimately more impactful.

2. **Affirmations:** Embracing positive affirmations can be a powerful tool to channel your thoughts towards your desired outcome. By repeating statements that align with your goal, such as "I am successful. I attract abundance" or "I am healthy," you can strengthen your belief that your goal is already within reach.

3. **Mindfulness:** Practice mindfulness by staying present in the moment and observing your thoughts without judgement. This will help quiet your mind and enhance your ability to focus on your desired outcome. By being mindful, you can prevent negative thoughts and doubts from interfering with your manifestation efforts.

4. **Gratitude:** Embracing a mindset of gratitude can have a profound impact on your life. When you acknowledge and appreciate the things you already have, you create a positive energy that attracts even more positivity. By shifting your focus to gratitude, you open yourself up to a world of positive experiences.

5. **Meditation:** Incorporating regular meditation into your routine can enhance your connection with the quantum field

and improve your ability to concentrate. During meditation, imagine your desired outcome and experience the emotions linked to its realisation. This practice reinforces the Observer Effect and boosts the chances of manifesting your desires.

Conclusion: Embracing the Observer Effect

The Observer Effect goes beyond mere scientific curiosity; it holds the potential to be a transformative force in our lives. By grasping and using this principle, we can gain mastery over our reality and mould it to align with our deepest desires. Whether we employ visualisation, affirmations, or other manifestation techniques, the crucial element is to concentrate our thoughts and intentions with resolute clarity and unwavering belief.

As I look back on my personal journey, from moments of deep thought by the ocean to my adventures in law enforcement, I am reminded of the powerful influence that careful observation has had on my life. It has been the catalyst for my achievements, the source of stability during uncertain times, and the compass that has guided me to my current position.

In the upcoming chapters, we will delve into additional quantum principles like superposition and entanglement and how they impact manifestation. However, for the time being, let's embrace the Observer Effect's potential and use it to manifest the reality we yearn for. Both quantum mechanics and spiritual teachings propose we possess the ability to shape our world from within, eagerly awaiting our release.

References

1. Bohr, Niels. *Atomic Physics and Human Knowledge*. New York: John Wiley & Sons, 1958.
2. Heisenberg, Werner. *Physics and Philosophy: The Revolution in Modern Science*. New York: Harper & Row, 1958.
3. Goddard, Neville. *The Power of Awareness*. New York: Penguin Books, 1952.

4. Young, Thomas. *Lectures on Natural Philosophy and the Mechanical Arts.* London: Longman, Hurst, Rees, and Orme, 1807.

5. Penrose, Roger. *The Road to Reality: A Complete Guide to the Laws of the Universe.* London: Jonathan Cape, 2004.

CHAPTER 5

Superposition and Possibility – Expanding Potential

"Reality is just a mere illusion, but boy, is it persistent!"
"Imagination is more important than knowledge."

– Albert Einstein

Introduction: A Delightful Journey into Endless Possibilities

Picture yourself standing at the edge of a boundless field, stretching as far as the eye can see. Every step you take reveals a multitude of pathways, each leading to a unique destination. This image beautifully captures the essence of a concept that is central to quantum mechanics—superposition. It is both simple and profound, inviting us to delve into the depths of this fascinating field.

The principle of superposition states that particles have the ability to exist in multiple states at the same time until someone observes them. In the realm of quantum physics, this concept pushes the boundaries of our conventional understanding of reality. It implies that multiple possibilities coexist until we decide, which then merges all other possibilities into a single reality. For some, this concept may appear abstract, but for the process of manifestation, it reveals a boundless realm of possibilities.

By embracing the principle of superposition and aligning it with the Law of Assumption, we can open ourselves up to a world of endless possibilities. By holding multiple potential outcomes in mind, we can shape our reality and bring about the desired manifestations. In this chapter, we will delve into the captivating interplay between superposition and manifestation, using real-life examples, scientific explanations, and philosophical insights to breathe life into these concepts.

The Science of Superposition: Exploring All Possibilities Simultaneously

The fundamental concept of quantum mechanics revolves around the notion that particles, like electrons, do not possess a singular, predetermined state. Instead, they exist in a superposition of all conceivable states. Until someone makes an observation, a particle has the fascinating ability to exist in multiple places at once, spin in both directions simultaneously, or exist in a combination of states.

Superposition in Daily Life: The Power of Choice

Although we often associate superposition with the realm of quantum particles, we can feel its influence in our daily lives. It's fascinating to think that our lives are brimming with endless possibilities, just waiting to be brought to life through the choices we make.

Let's look at the life of Thomas Edison, the brilliant inventor behind the phonograph, the motion picture camera, and the electric light bulb. Edison embarked on a journey filled with endless exploration, often encountering setbacks and challenges. He once famously said, "I have not failed. I've just found 10,000 ways that won't work." This mindset of viewing failure as a stepping stone towards success reflects the principle of superposition.

Edison's unwavering determination to explore various possibilities, despite encountering many setbacks, enabled him to transform a multitude of potential outcomes into one triumphant reality. His remarkable talent for considering multiple potential outcomes instead of becoming discouraged by initial setbacks exemplifies the power of embracing possibilities to unlock one's full potential.

Embracing Multiple Possibilities: The Path to Manifestation

Embracing multiple possibilities is key in the process of manifestation, as superposition teaches us. Rather than fixating on a single outcome, it's important to remain open to different potentialities. Being open to different paths that can lead us to our desired goal is essential, rather than being indecisive or scattered about our intentions. By adopting this approach, we enhance our chances of attaining our objective as we embrace a multitude of possibilities.

Let's consider the story of Dhirubhai Ambani, the Indian business tycoon who founded Reliance Industries, one of the largest conglomerates in India. Ambani's incredible journey from humble beginnings to becoming one of the most successful businessmen in the world is a true testament to the boundless potential that lies within us all.

Ambani began his journey as a clerk in a trading firm in Yemen, fuelled by ambitious aspirations of building a formidable business empire. He explored various industries and business models, refusing to be confined to just one. Instead, he ventured into different fields, ranging from textiles to petrochemicals to telecommunications. His ability to embrace various possibilities enabled him to overcome obstacles and capitalise on opportunities, ultimately resulting in the establishment of Reliance Industries.

Ambani's story beautifully exemplifies the concept of superposition in action. Through his open-mindedness and willingness to explore distinct possibilities, he could bring his vision of a thriving business empire to life.

Superposition and the Law of Assumption: Unleashing the Power of Imagination

The concept of superposition is more than just a theoretical idea; it is put into practice by many accomplished individuals to achieve extraordinary outcomes. Here are a few examples of how superposition has been used in various fields:

1. **Indra Nooyi:** The former CEO of PepsiCo exemplifies the power of embracing multiple possibilities and how it can pave the way for groundbreaking success. Her visionary approach marked Nooyi's leadership style, allowing her to expect and overcome challenges and steer PepsiCo towards success in a variety of markets. She had a broad vision and explored various possibilities for growth, such as expanding the company's portfolio to include healthier products. Her exceptional skill in tackling intricate challenges while considering multiple perspectives played a pivotal role in propelling PepsiCo to become a dominant force on the global stage.

2. **Ratan Tata:** Another notable figure from India is Ratan Tata, who served as the chairman of Tata Group. Under Tata's visionary leadership, the company ventured into diverse industries such as steel, automobiles, and telecommunications, marking a significant expansion. Tata's open-mindedness and eagerness to venture into new territories and embrace various opportunities enabled the company to expand and diversify, establishing itself as one of India's most prosperous conglomerates. His vision of creating the Tata Nano, the world's cheapest car, reflected his strong belief in making transportation affordable for millions of people.

3. **Sheryl Sandberg:** Many people widely recognise Sheryl Sandberg, the COO of Facebook, for her remarkable resilience and exceptional ability to navigate through various career opportunities. Sandberg's career journey was anything but straightforward. She ventured into different fields like government, international development, and the tech industry before eventually finding her place at Facebook. Her open-mindedness and ability to navigate through various situations have positioned her as a highly influential figure in the world of technology.

The Importance of Visualisation in Expanding Possibilities

Visualisation is an incredibly powerful tool that allows us to tap into the magic of superposition in the manifestation process. By envisioning various potential outcomes, we can broaden the range of possibilities at our disposal and enhance our chances of attaining our desired objectives.

Take a moment to reflect on the incredible achievements of **Michael Phelps,** the unparalleled Olympian who has amassed an astounding 23 gold medals. Phelps is renowned for his intense training routine, but what many may not realise is the crucial role visualisation played in his remarkable achievements. Phelps would imagine all the different scenarios that could happen during his races, from the best possible swim to potential obstacles, like his goggles getting filled with water. Through this practice, he could cultivate a mindset that enabled him to adapt and excel, regardless of the circumstances that unfolded during the race.

Phelps's adept use of visualisation shows the practical application of the principle of superposition in real life. Through his imaginative thinking, he skilfully expected and achieved the result he aimed for—triumph in the pool.

Superposition and Decision-Making: Navigating Life's Choices

The principle of superposition can be incredibly useful for decision-making. Life is a tapestry of choices, each one holding the power to shape our destiny. By recognising the multitude of possibilities that coexist, we can make better decisions that are informed and empowering.

Consider the case of Jeff Bezos, the visionary behind Amazon. Prior to launching Amazon, Bezos had a thriving career at a successful hedge fund in New York. He found himself at a crossroads, torn between the security of his stable and lucrative career and the allure of taking a leap of faith and starting an online bookstore.

Instead of viewing it as a simple either/or decision, Bezos wholeheartedly embraced the concept of exploring multiple possibilities. He carefully weighed the outcomes, considering both the positive and negative aspects, before reaching a decision. He approached his choice with a forward-thinking mindset, asking himself, "Will I look back in 30 years and wish I had given this a try?" This mindset enabled him to make a courageous choice that ultimately resulted in the establishment of one of the most prosperous companies globally.

Bezos's decision-making process reflects the principle of superposition. Through his thoughtful consideration of various possibilities and his ability to stay focused on his long-term vision, he skilfully manoeuvred through uncertain circumstances and successfully brought his goals to life.

Practical Techniques for Embracing Superposition in Manifestation

Now that we've delved into the concept of superposition and its connection to the Law of Assumption let's explore some practical techniques that can help you embrace multiple possibilities in your manifestation practice:

1. **Mind Mapping:** Generate a mind map of your objectives and the different avenues that might lead to their fulfilment. Immerse yourself in a world of endless possibilities and embrace the beauty of diverse outcomes.

2. **Scenario Planning:** Just like visualisation, scenario planning involves envisioning various possibilities that could unfold as you work towards your goals. This approach enables you to mentally anticipate different scenarios and adjust your actions accordingly.

3. **Positive Affirmations:** Embrace the power of positive thinking and open yourself up to endless possibilities. For instance, "I am receptive to the endless possibilities that life presents," or "I welcome the potential for success in diverse ways."

These positive statements can help change your perspective from a rigid outcome to one that is receptive to a multitude of opportunities.

4. **Gratitude Practice:** Embrace a mindset of appreciation for all the possibilities that come your way, even the ones that may seem difficult or surprising. By embracing all outcomes with gratitude, you cultivate a mindset that radiates positivity and invites an abundance of opportunities into your life.

5. **Adaptability:** Be open to different approaches when working towards your goals. Stay open-minded and flexible, understanding that being adaptable and willing to change course when needed is key to achieving success.

Final Thoughts: Embracing the Limitless Potential of Superposition

The principle of superposition provides a valuable framework for exploring new possibilities in the manifestation process. By recognising the infinite possibilities that exist and embracing the power of assumption, we can shape a reality that aligns with our greatest dreams and desires.

As you delve deeper into this book, I invite you to explore these ideas in your own life. Embrace the notion that endless possibilities await you and use the strategies outlined in this chapter to navigate the intricacies and uncertainties of life's decisions.

Always remember that you can shape your own reality. By understanding the power of superposition and the Law of Assumption, you can tap into your mind's incredible potential and create a life filled with endless possibilities.

Sources

1. Feynman, Richard P. *QED: The Strange Theory of Light and Matter.* Princeton University Press, 1985.

2. Nooyi, Indra. *My Life in Full: Work, Family, and Our Future.* Penguin Press, 2021.

3. Tata, Ratan. *The Wit & Wisdom of Ratan Tata.* Rupa Publications India, 2019.

4. Sandberg, Sheryl. *Lean In: Women, Work, and the Will to Lead.* Knopf, 2013.

5. Bezos, Jeff. *Invent and Wander: The Collected Writings of Jeff Bezos.* Harvard Business Review Press, 2020.

CHAPTER 6

Quantum Entanglement and Connectedness: Moving Beyond Duality in Manifestation

"We are like islands in the sea, separate on the surface but connected in the deep."

– William James

In the tranquil village of Coonoor, beneath the shade of a huge banyan tree, one can barely hear the distant rustle of pine needles and the light chatter of people below. According to local legend, this tree has silently witnessed countless stories of love, loss, and connection.

The remarkable story of Arjun and Meera stands out; they were best friends from childhood and could share each other's suffering even when separated by oceans. Many people thought this peculiar pattern proved the existence of invisible connections between all living things, connections similar to the mysterious quantum entanglement.

Did your intuition ever turn out to be right when it warned you that someone important to you was in danger? On the flip side, you may have experienced a remarkable coincidence, such as reuniting with a long-lost acquaintance via a phone call after reminiscing about him. The usual explanation for such occurrences is randomness, but the enigmatic phenomenon of quantum entanglement may reveal a hidden reality.

Envision two butterflies gracefully circling each other in a bright field. One appears to be fluttering westward and the other eastward, suggesting that their paths are diverging. Unbeknownst to them, a strange force binds them together. Regardless of the distance between the two butterflies, observing one will cause the other to rapidly change its wing colour to match. This boils down to quantum entanglement. When two particles couple, their states are immediately influenced by each other, regardless of how far apart they are. Their apparent coexistence begs the question of our standard spatial conceptions.

The Concept of Quantum Entanglement: A Primer

Quantum entanglement is a fascinating and perplexing phenomenon in the field of quantum mechanics. When particles are entangled, their states are interdependent, and the motion of one particle has an instantaneous effect on the other, regardless of how far apart they are. Imagine two electrons linked in a way that defies the laws of space and time. No matter how far apart they are, the two particles will immediately reflect each other's states if one of them changes. The results of this study challenge long-held assumptions about the universe's interconnectedness, suggesting instead a quantum level of connectivity.

One fine night, Dr. Michael—a famous surgeon—dreamt vividly that he was performing a complicated operation. He encountered an issue in the dream that he had never experienced before. The next morning, a patient was wheeled in as an emergency case with the same rare condition he had dreamt of. The surgeon saved the patient's life during the surgery. Remarkably, Dr. Michael found himself in the exact situation of his dream. Thanks to the dream, he knew exactly what to do. Dr. Michael is convinced there was some message, some preparation from the collective consciousness, and it found its way through his dreaming mind.

Embedded Substances' Mysterious Dance

It wasn't until the 1980s that scientists were able to confirm Albert Einstein's 1935 proposal of entanglement. He described the phenomenon as 'spooky action at a distance', highlighting how contradictory it seemed. Something about the concept that subatomic particles could interact with each other instantly and profoundly over great distances unsettled Einstein.

Subsequent research disproved his earlier assumptions about entanglement and uncovered an unexplainably high degree of interconnection. When two particles become entangled, their fates are effectively intertwined. They exist in the same quantum state, so any measurement on one will instantly affect the other, regardless of their distance.

A Lesson on Complexity from Neville Goddard

Neville Goddard emphasised the mind's capacity to manifest its own reality and the interconnectedness of all things. This view is consistent with quantum entanglement, which suggests that our thoughts and intentions can influence the cosmos in microscopic, imperceptible ways.

Being Is All There Is

Throughout his lectures on the unity of consciousness, Goddard emphasised that each of us is an interconnected part of a greater whole. Two particles' states are intrinsically linked, regardless of their physical separation, according to quantum entanglement, which is consistent with this theory. Our awareness is inextricably bound to the cosmic consciousness, much like entangled particles; as a result, we are both impacted by and able to shape our immediate surroundings.

There were countless instances when I sensed an invisible hand guiding me towards my goal of becoming an IPS officer, much like the link between entangled particles. I remember getting this strange urge to pay extra attention to one topic while I was preparing for the Indian

Police Service Examination; much to my surprise, that topic ended up being very important on the actual exam. I felt as though the universe was whispering in my ear, nudging me towards success. This unseen connection has become as reliable to me as quantum entanglement.

For example, think about the Maya and Priya twins. They have an uncanny knack for reading people's emotions. They often had comparable physical and mental experiences despite being born and brought up in different parts of the world. Maya had a severe anxiety attack one afternoon while on vacation in the mountains. A chill ran down her back, and her heart raced. Priya had a terrifying dream in which Maya was hanging precariously from a mountaintop. After slipping on a loose rock, Priya nearly reached out to Maya in response to the dream, but she managed to regain her balance. The proof pointed to a more profound relationship, perhaps analogous to quantum entanglement, than merely a strong twin tie.

Take into consideration Linda McCartney, the late wife of Sir Paul McCartney, before passing judgement on him. The famous musician Paul often spoke of the deep, almost mystical connection he shared with Linda. Not only were they best friends for life, but there was an indescribable harmony between them as well.

Paul spent a single evening in 1973 secluded in a recording studio, composing music. With great difficulty, he sought out the perfect tune for this session. Linda was engrossed in her reading nook across the nation. Despite her infrequent solo piano practice, she felt an overwhelming need to play that piece alone. Out of nowhere, a tune came to her, and she started to play it. Paul had spent a considerable amount of time searching the studio for one that was similar. Paul called her because of some strange attraction, and while they were on the phone, Linda sang this song to him. It wasn't just a creative coincidence; it proved the deep quantum bond between two harmonious souls.

But consider the friendship that developed between the revered Indian sage Swami Vivekananda and his mentor, Ramakrishna Paramahamsa. Vivekananda often discussed the profound intellectual and spiritual bond that he shared with Ramakrishna. Vivekananda often felt his late guru Ramakrishna's guidance because of the close relationship they shared. Upon his return to India from his Western travels, Vivekananda found that the periods of deep insight and problem-solving had really occurred during the particular prayers and meditations done in memory of Ramakrishna. This deep and eternal connection personified the idea of quantum entanglement, which transcends the limitations of space and time.

Despite featuring different protagonists and locations, these stories share a central theme: the enigmatic connections that bind people together. Similar to entangled particles in quantum physics, we can overcome the constraints of time and space in our interpersonal relationships. Not only can spiritual guidance and music transcend national boundaries and oceans, but so can psychic readings. In such cases, the principles of quantum entanglement are applicable.

Realisation and entanglement go hand in hand. Because our intentions and thoughts can affect entangled particles across enormous distances, it is possible that life's events are more predetermined than they appear. Imagine yourself as a lustrous thread delicately interwoven into the fabric of existence.

Maria and Lucia were two girls born and raised together in a tiny town in Italy. Over time, they developed such a strong relationship that it sometimes went beyond the usual understanding of the world. As grown women, they were now separated by miles but often had similar dreams.

Once, Maria saw a hill with a single beautiful, ancient olive tree standing alone on top of it, its branches slowly moving in the breeze. She got a call from Lucia, who described the same tree that had been in her dream. The next morning, they both awoke with the same image in their heads. They had never visited that place. The common image in their minds was definitely not a coincidence. It was more a sign of their deep, entangled connection—the way quantum particles maintain the phenomenon of being connected despite distance. It seems as though a rich tapestry of feelings and thoughts is transmitted irrespective of distance and time.

Focusing on a single object causes its hue to intensify, which in turn affects the entire universe. Through the mysterious dance of entanglement, you can align your focused intention with frequencies similar to your goal, attracting opportunities and experiences that help you reach it.

There was James, an artist in New York, who couldn't help but think of giving up on his art. One rainy afternoon, feeling very discouraged, he said to himself, "I will take a walk and look around the city."

As he turned into a little, out-of-the-way street, he came across an old art store, and in the window, with no other picture around it, hung a painting that was an almost exact reproduction of a long-cherished fancy of his, one that figured repeatedly in his many sketches. The uncanny chance of this occurrence banished his discouragement as though by magic and, in this way, helped him suppress his reasons for wanting to give up. Years later, when he became successful, James remembered this moment as one of the most striking coincidences, as though the universe itself had given him a little push at the right time.

The fascinating possibilities of interconnected consciousness are only now being discovered by scientists. Dr Roger Nelson of Princeton University heads the Global Consciousness Project (GCP), which spans the globe and seeks to understand interconnected consciousness. To find patterns in apparently random data during major global events, the GCP uses a decentralised network of Random number generators (RNG). The GCP has demonstrated some deviations from randomness in situations involving the emotions and psychology of a large number of people, such as natural disasters, terrorist attacks, and huge cultural festivals. Taken together, these findings point to a more entangled reality that satisfies the entanglement criteria, wherein the collective intentions and emotions of humans could influence the behaviour of RNGs.

The implications of entanglement for manifestation are immense, even though research into it is still in its early stages. By embracing the interconnectedness demonstrated by quantum mechanics, we can move beyond an individualistic mode of manifestation and enter the realm of collective consciousness. We can strengthen our shared beliefs and potentially make a difference that affects more than just ourselves if we work together and stay committed to the same goals.

Imagine a symphony performed by a group of musicians, all of whom are exceptionally talented and trained. On their own, the tones might produce an unpleasant dissonance. But when they work together in harmony, it is like witnessing a spectacular performance that no one can ignore. When our desires are in harmony with the thoughts of the collective consciousness, it feels as though our lives and the world around us are orchestrating a symphony of manifestation.

Although entanglement is purely theoretical, it does have some intriguing real-world applications. The following examples demonstrate the profound effect of interconnection:

Once weekly in a thriving West Coast city, there is a space called the healing circle where people can come to meditate and heal. They believe that healing and transformation are possible when they focus their collective intentions on a specific person or situation. Many incredible transformations and recoveries have occurred over the years, and they credit the power of their linked consciousness for these changes.

One of the most moving cases was Sarah, a young woman fighting a debilitating illness. Sarah was the centre of the group's healing efforts, and they envisioned her fully recovered and enveloped in a golden light. Within a few weeks, Sarah's condition began to improve, much to the doctors' surprise. Such occurrences have only served to deepen my belief in the power of focused attention and its capacity to shape our surroundings in line with entanglement principles.

When was the last time you effortlessly got back in touch with a long-lost friend? These kinds of events are called synchronistic rencontres. Perhaps you received a call while fantasising about a potential new employment opportunity. What seems like discordant data may actually be directions from the entangled field.

Behind the Scenes Link

Exactly how humans relate to non-human creatures isn't always easy to put into words. We can all relate to the experience of being a pet owner whose animal displayed some kind of seemingly telepathic communication—be it a knowing look, a courteous prod, or something else entirely. The precise nature of this connection is still debatable, but some researchers have proposed that it may represent a weak form of entanglement. Animals may be capable of picking up on human emotions and thoughts via the entanglement field due to their heightened sensitivity to even the most minute changes in energy.

Given the depth of her connection with her horse, Emily exemplifies this phenomenon perfectly. A shift in her horse's demeanour signalled to Emily that she was about to experience a metamorphosis in her own feelings. She had to get a handle on her emotions before she could influence her horse's actions, as her emotions dictated her horse's behaviour. Beyond the typical avenues of communication, this mirroring action seems to suggest a connection analogous to entanglement.

Methods for Physically Verifying the Connection Between Mindfulness and Entanglement

Although entanglement is still in its infancy, scientists are rapidly providing evidence that it does, in fact, exist. This phenomenon is illuminated by these important studies:

- In 1982, Alain Aspect conducted a groundbreaking experiment called 'The Aspect Experiment', which proved that entangled particles can influence each other rapidly despite their physical separation. This finding provided evidence of a deeper, non-local connection across the cosmos and challenged the conventional idea of location. The Aspect Experiment and subsequent research provided indisputable evidence of quantum entanglement, enabling new understandings of connections.

- A new entanglement issue was demonstrated in the delayed-choice experiment in 2007 by researchers from the University of Vienna. It is possible to observe the entangled state of particles even after they are separated. This lends credence to the idea that consciousness could play a role in the collapsing of wave functions due to their quantum mechanical basis. By suggesting a potential link between awareness and the mixed field, the delayed-choice experiment supports Neville Goddard's assertions regarding the effectiveness of directed thought.

How does consciousness interact with the faint light that all living things emit? That is the question that biophotonics aims to answer. Biophotons, which some researchers believe aid cellular communication, could be part of a biofield that connects all living things. Though still in its early stages, the study has already uncovered intriguing clues regarding the connection between human awareness and the very nature of relationships. Research into biophotons may shed light on how our thoughts and intentions influence the entangled field, which could lead to further proof of the principles of quantum manifestation.

A Holistic Perspective on Being

Quantum entanglement exemplifies the interconnectedness of everything in the cosmos. By understanding the interdependence of all things, we can harness the power of entanglement to create the utopia we envision. Among the numerous significant takeaways from our investigation are:

1. No one truly exists in isolation. Everyone is dependent on and interdependent with every other person in the vast, unseen network of reality. Because entanglement shows how

interconnected everything is, our thoughts, emotions, and actions can change the world in profound ways.

2. A group's power is greater than the sum of its parts. We wanted to show that we were all in this together, so we formed the healing circle. A greater impact than the sum of its parts is possible when individuals combine their efforts to achieve a shared goal. The idea that the combined state of entangled particles can affect the overall outcome is consistent with quantum entanglement. When our goals are in harmony with those of people who share our values, we can tap into a greater power for manifestation.

There was a group of villagers who used to get together once a month in someone's home or, in pleasant weather, at a nearby shrine. In the small community in Japan, for any type of collective intention, they had developed a practice they referred to as 'harmony gathering'. They believed that combining their intentions could facilitate positive change, not only in their lives but also for the entire village. Through the years, they attributed the village's good fortune and resilience to these gatherings.

To come together with a common intention was, for them, a way to entangle individual desires toward weaving a fabric of collective reality, much like the quantum mechanics of entangled particles.

Synchronicities and What Could Be Behind Them Besides Just Coincidences

3. This is something that must be known. Increasing our sensitivity to synchronicities and subtle clues will help us navigate the entangled field more effectively. Be alert for dreams, symbols, or encounters that appear to be completely arbitrary. These clues from the intricate web may, with diligence and focus, lead you to your desired outcomes. You have to pay close attention if you want to tap into the information of the related fields.

"Synchronicity is an ever-present reality for those who have eyes to see."

– Carl Jung

Sarah, a mum with two little ones, had a grandma's locket—a family heirloom—that had gone missing. She looked everywhere and, as the days went by, grew desperate. Then, one fine day at the store, a stranger bumped into her with such force that he dropped his wallet. As she bent down to help him gather his belongings, something shiny caught her eye under the counter. There, dusty but unmistakable, lay her lost locket.

The man, noticing her surprise, smiled and said, "Sometimes things find us when we stop looking." For Sarah, that moment was more than a coincidence. It reminded her that the universe has a way of bringing us what we need at the perfect time, even when we're not searching for it.

4. **Getting Things Done Becomes Much Easier with a Clear Goal in Mind.**

    Despite the strong link between ideas and reality that entanglement suggests, taking action is still crucial. Our focused intentions light the manifestation fuse, but it is our concerted actions that truly ignite the fire. Demonstrating your commitment to your goals through creative action accelerates the manifestation process by bringing harmony to both your inner and outer worlds.

5. **Expressing Appreciation Strengthens the Bond.**

    You can maximise the results of your manifestation efforts by maintaining an attitude of deep gratitude. Gratefulness is a powerful energy practice that can lift your spirits and help you attract more of what you want into your life. Gratitude strengthens the connection between your determination and the infinite universe of possibilities, making it easier to turn your dreams into reality.

Practical Applications of Quantum Manifestation Methods: The Entangled Web

In previous chapters, we explored the fascinating world of quantum entanglement and how it may affect manifestation. All realities are interdependent, and we demonstrated this by precisely manipulating the entangled field. The moment has come to implement these principles. This chapter lays out a variety of methods for using entanglement to bring your dreams into reality.

1. Relaxation Techniques for Entanglements

By establishing a profound connection to the complex web through this guided meditation, you can release your dreams into the limitless field of opportunity.

- Find a quiet place where you can settle in and adopt a comfortable position. Place your palms on your heart and take a few slow, deep breaths to calm yourself.

Imagine yourself as a unique and brilliant thread entwined with all the others in the tapestry of existence. A beacon for your thoughts and emotions, this thread throbs with your awareness.

Consider instead a vast tapestry that connects you to all other forms of life in the cosmos. Connecting with the field is done this way: Let yourself be carried along by the boundless energy of possibility— the vast network of interconnections.

Focusing on a specific objective is essential when planting seeds. As you imagine yourself having already achieved it, you will feel a sense of accomplishment. Allow this mental image to fill your entire being, enabling the thread to represent your mission.

Envision your desire's thread intricately interwoven with the fabric, harmonising with other strands that resonate with the same frequency. Picture these bonds deepening, the volume of your inner voice increasing, and blessings and opportunities flooding your path.

A key component of the integration technique is returning to a regular practice of deep breathing. Surrender to the euphoric energy—a fleeting expression of your link to the entangled field—and allow yourself to be carried away. Take charge of the day and make an impression that will be hard to shake.

2. The Team That Establishes Goals

Incorporating the power of group focus and entanglement principles, this practice enhances the manifestation process.

Seek out and spend time with individuals who share your interest in quantum manifestation. Use this to assemble your tribe. This could take the form of a local manifestation circle, a social circle, or even a family reunion. Intentional centering can be aided by creating a sacred space. Set the mood with calming music, candles, and a comfortable circle of people.

Everyone describes their perfect life, creates a visual representation of it, and shares the emotions that go along with it as part of the common goals activity. Everyone needs to be involved and ready to help.

It's important to come together as a team after everyone has had a chance to speak. To visualise the strength of your combined goals, picture a brilliant beam of light shining in the middle of the circle. Imagine this beam of light expanding, merging with the entangled field, and opening doors to opportunities that benefit everyone involved.

Have faith that the entangled field is working its magic as you tenderly send your wishes out into the universe, and be grateful for the chance to work together.

3. Synchronicity Journal

You can master the entangled field's subtle signals and use them to your advantage with practice.

- **Daily Devotion:** Set aside a portion of your journal or diary to keep track of synchronicities. Jot down every coincidence that occurs, no matter how insignificant it may seem. Examples include overhearing a conversation that confirms your life goals or noticing a recurring symbol.

Remember that the field of entanglement is sending you messages through seemingly unconnected events. Analyse your past experiences in light of your current goals.

Even if a synchronicity appears small, an actor shouldn't ignore it. Accept it with grace as a gentle reminder to proceed correctly. Utilise your drive to achieve your objectives and make the most of the fresh opportunities it presents.

4. Enabling Biofeedback

By combining your biofeedback system with the energy of purpose, this method strengthens the connection to the entanglement field.

Before attempting to connect with the biofeedback loop, it's a good idea to calm your body with deep breathing or progressive muscle relaxation exercises.

Creating a Mental Picture of Yourself Reaching Your Goals

Visualise exactly what you want to happen. Surrender to the euphoria of finally achieving it.

Focus on how your body feels and give yourself permission to experience positive emotions as you strive to be more in tune with them. Be alert for changes in temperature or numbness anywhere in your body.

When you become aware of these sensations, intentionally amplify them. As they emerge from your centre and become entangled, picture them growing stronger. You can improve your chances of manifesting by using this biofeedback loop to amplify your intention within the field.

5. A Method of Expressing Gratitude

A sense of oneness with the universe can be enhanced by expressing gratitude for what you have. Paying attention to and being thankful for the good things happening in your life is a great way to attract more of those things into your life.

Establishing a Practice of Gratitude

Every day, preferably first thing in the morning, jot down at least one thing you're grateful for. Whether it's a sound mind, a loving family, a beautiful morning, or something as simple as a cup of coffee, take stock of your blessings and give thanks.

Expressing gratitude might help put yourself in the receiver's position. Feel every bit of joy that comes from being blessed abundantly.

Imagine a lake reflecting your appreciation as you practise intertwined gratitude. The entangled field is a powerful amplifier, so let your positive energy merge with it. By aligning your energy with the frequency of your desires, you can attract more abundance into your life.

Be Consistent All the Way Through!

The strategies previously discussed are most effective when practised regularly. The life-altering effects of incorporating entanglement field awareness and focused purpose into your routine will become immediately apparent.

Here Are a Few More Things to Consider as You Work to Make Your Dreams Come True

Follow Your Joy: Doing what brings you joy and allowing yourself the freedom to pursue your dreams is what life is all about. Possessing a magnetic force that attracts opportunities becomes your reality when you live your life in alignment with your deepest desires.

Let Go of Resistance: To overcome resistance, it is essential to release self-deprecating ideas and limiting beliefs. By following these steps and having faith in the law of attraction, you can forgive both yourself and those around you. Refusing to let go of resistance prevents abundance from entering your life.

Take Inspired Action: Instead of doing nothing and hoping for the best, start working on creative endeavours. Don't give up on your goals—progress, however small, is still progress. Taking action shows the universe that you're committed, which could speed up the manifestation process.

Celebrate Small Wins: No matter how small, celebrating your successes can keep you motivated and boost your faith in your ability to make your dreams a reality.

Dispelling Common Myths and Overcoming Common Difficulties

You should be mindful of and strive to overcome misconceptions and obstacles, even though quantum entanglement offers powerful manifestation tools.

Scepticism and Uncertainty

Many people are understandably suspicious of quantum entanglement and connectivity because of their mystifying features. To triumph over these obstacles, you must find scholarly research and instances that support these claims. When you band together with people who share your passion, you can foster an atmosphere where innovative ideas can thrive.

Partitioning and Staying Away From

It might be difficult to embrace the ideas of entanglement when you're feeling lonely and disconnected. To alleviate these feelings, try attending community events, joining an intention-setting circle, or meditating in a group. The most effective way to harness the potential of entanglement and collective consciousness is to fortify your relationships with those around you.

Concept of Fate and Free Will

Picture yourself as if your decisions are being guided not from inside you but by an invisible network of connections—it is almost as if a cosmic web led you to it.

Every thought, every choice, and every interaction is part of a grand design, yet within this design, you still possess the power of free will.

Can it be that free will and destiny aren't competing forces, but both wear different masks to play out as a team—in perfect harmony, doing the most beguiling dance of random quantum entanglement?

It was in the 1920s, in the heart of Paris, one evening when a young writer, Isabelle, entered a smoky, low-lit café, her mind besieged by self-doubt.

A kind elderly stranger approached her while she was sitting and said, "I feel compelled to tell you something: never stop writing." Years later, after publishing a novel that would become a beacon of hope for many, Isabelle often reflected on that night and the mysterious stranger, believing it was a moment of divine intervention—a nudge from the universe guiding her along her path.

To Wrap Things Up

The concept of quantum entanglement offers a fresh perspective on the cosmos and our role within it. By embracing the concepts of quantum entanglement and putting the methods outlined in this chapter into practice, you can unlock a door to manifestation that will change your life. Integrating your aspirations with the fabric of reality transforms you into a dynamic participant in the expansive realm of existence. If you can bring yourself into harmony with the universe's complex web and keep your focus, you can access its infinite potential. Keep an open mind and be receptive to the messages from the entangled field; as you go along this path, you will witness the magic of quantum manifestation unfold.

To assist you in quantum materialisation, the following chapters will offer further scientific explanations, examples, and instructions for applying the principles. Let us embark on a journey of limitless potential, where our dreams and ambitions dictate our destiny, and embrace the fundamental interconnectedness of all things.

"There are more things in heaven and earth, Horatio, than are dreamt of in your philosophy."

—William Shakespeare, *Hamlet*

The fascinating entanglement framework provides a view on manifestation, even though the precise ramifications of this concept are still the subject of vigorous academic debate. Additional research is necessary to understand entanglement and its potential effects on our reality.

CHAPTER 7

The Quantum Power of Intention – Shaping Reality with Thought

It was a gentle morning in Chennai, the sunlight spilling across the quiet streets, casting long shadows on the waking city. As the world stirred into motion in these early hours, it felt as though each slight movement—the rustling leaves, the soft footsteps, the hum of distant voices—was part of an intricate, interconnected rhythm. It was a time when one could pause, observe, and sense that beyond what can be seen, life was woven together by invisible threads, each one vibrating with potential.

Such mornings have often made people reflect on the power of manifestation. Manifestation might seem a distant concept, perhaps abstract or mystical to the untrained mind. But within the pages of quantum creation, a deeper understanding unfolded. The ability to manifest was rooted not in magic or wishful thinking but in the very fabric of quantum reality.

In previous chapters, concepts like the double-slit experiment, superposition, and entanglement were explored, each offering a glimpse into the hidden ways our intentions interact with the universe. As the city began its day, it was time to revisit these ideas—to see them anew and understand how they underpinned the practice of manifestation.

1. The Quantum Wave of Intention

In quantum physics, every intention can be likened to a wave of potential, much like the behaviour observed in the double-slit experiment discussed earlier. In that experiment, particles exhibited wave-like behaviour, existing as probabilities rather than fixed entities. Only when observed did they 'collapse' into a definitive state.

This phenomenon is a beautiful analogy for manifestation: our desires and goals reside in a field of possibility, waiting to be brought into reality through focused attention.

As the day unfolded, it became clear that intention was more than just a thought; it was a wave of energy, much like those particles from the experiment. When one focused on an intention, it was as though they were 'observing' this desired outcome, guiding it from a field of possibilities into the realm of reality. Just as light can shift between wave and particle, desires can also change from thought into experience.

2. Entanglement: The Threads of Connection

The concept of entanglement, discussed earlier, reveals a fascinating dimension of quantum creation. Entanglement shows how two particles, once connected, remain linked across distances, their states instantly reflecting changes in one another. This idea, often described as 'spooky action at a distance', forms a vital piece of the manifestation puzzle.

When someone holds an intention with deep emotional resonance, it's as if they create an entangled bond between themselves and their desired outcome. Through this connection, the intention resonates through the quantum fabric, seeking to bring that possibility closer.

It became easy to imagine this interconnectedness in the stillness of early morning reflections, like an invisible thread stretching between one's inner desires and the vast, unfolding universe.

In this quiet space, focusing on a desire became a powerful act, an entangled connection with what was to come. Holding that intention without wavering strengthens the bond, drawing the outcome closer. It is a process of nurturing, allowing, and releasing, knowing these threads are woven in a realm beyond time and space.

3. Embedding Intentions in the Subconscious: The Influence of Repetition

In Chapter 12, we will read how our subconscious is a embedded repository of patterns and beliefs that shape our reality. Neuroscience shows that repeated thoughts and intentions mould the brain, a concept known as neuroplasticity. In quantum creation, embedding an intention is akin to embedding a frequency within one's being, aligning oneself with that possibility.

We align ourselves with the desired reality by training the subconscious through repeated visualisation and affirmation. As we discussed, this repeated focus is like observing a potential outcome repeatedly until it becomes our reality. Just as neurons wire together through repeated actions, each repetition becomes another brushstroke on the canvas of reality, embedding intention deeply within both the mind and energy fields.

As the day brightened, quiet reflection and focus became a form of intention-setting, a way to align with one's goals gradually. Each morning, through the simple practice of visualising and affirming, people imprint their desires within, allowing the subconscious to hold and amplify them.

4. The Double-Slit Experiment: Choosing Reality Through Observation

The double-slit experiment has been our guide for understanding the power of observation, showing how particles behave differently when observed. In the realm of quantum creation, our desires, like these particles, are waves of potential until we bring our attention to them. When we focus, we get these waves into a fixed state, shaping them into reality.

Here lies the beauty of manifestation: it is not about creating something from nothing but about choosing, through focus, the potential outcome we wish to see. Every thought and intention is like a gentle nudge, guiding the probabilities in our favour.

Reflecting on the quiet streets of Chennai, it became clear that reality was constantly shifting, always in motion. Each choice of focus was like a subtle adjustment, guiding one's path in unseen ways. In every moment, one could be the observer, the selector of possibilities, collapsing a desired reality into being through focused attention.

5. Quantum Superposition: Infinite Paths, One Chosen

The concept of superposition, previously discussed, suggests that all possibilities exist simultaneously until one is chosen. Quantum creation beautifully applies this idea: our lives are filled with infinite potential paths, each waiting to be realised. Just as particles can exist in multiple states simultaneously, our future holds countless versions of ourselves. The one we choose to focus on is the one that emerges.

In manifestation, to embrace superposition is to recognise that any reality we desire is already present, waiting for us to choose it. Each intention acts as a choice from this field of possibilities.

As people contemplate their desires, they realise they are not creating a future from scratch but aligning with an existing version of reality. By focusing on their intentions, they bring that version into clearer view, allowing it to materialise in their experience. It is a dance of possibilities, where focus and choice become the instruments of creation.

6. Aligning with the Frequency of Desire

Energy resonates at specific frequencies in the quantum realm, and our thoughts and emotions are no different. To manifest, we must align our frequency with that of our desires. This is not mere wishful thinking; it is resonance. Energy attracts energy of a similar frequency, and by aligning our emotions with our goals, we create harmony between our inner world and outer reality.

An abundant mindset is essential for those who seek abundance; for those who seek love, a loving mindset must prevail. As particles resonate at specific frequencies, our desires resonate at the frequencies we cultivate within ourselves.

On such a morning, embracing gratitude, joy, or love can become a practice of aligning with one's goals. By embodying the emotions of one's desired outcome, we harmonise our vibrations with the quantum field, fostering a seamless flow between inner intention and outer experience.

7. Releasing Attachment to Outcomes

One of the paradoxes of quantum creation is the importance of releasing attachment, a concept intertwined with our understanding of particle behaviour. Just as particles exhibit the greatest potential when not fixed in place, our desires flourish when we allow them the freedom to unfold. In quantum terms, when we release our tight grip on a goal, we enable it to manifest naturally without the interference of rigid expectations.

As one walks through the unfolding day, one is reminded to set desires free, trusting that the universe has a way of shaping them. Letting go is not a lack of belief but a sign of trust in the process, a surrender to the dance of energy and intention. In doing so, one aligns more deeply with the quantum flow of creation, allowing the manifestation to take its natural course.

In these quiet moments of reflection, the vastness of quantum creation becomes clear. Every thought, every desire, every belief is a thread in the infinite tapestry of reality. Manifestation is not an act of force but one of harmony, an alignment with the universe's rhythms.

In quantum creation, the path forward is not just about seeking; it's about being—a state of resonance, a gentle choice, a wave of intention that moves outward into the vastness. Each morning, with the city awakening and possibilities stirring, one can feel the universe hum with potential, whispering that creation is, and has always been, within reach.

And so, with every breath, every thought, one becomes a co-creator, painting the universe with the brush of intention, bringing forth the life imagined from the deep, boundless quantum field.

8. The Mirror of Quantum Entanglement: Reflecting Reality Within

Quantum entanglement reminds us that once particles become entangled at the subatomic level, their states remain linked across any distance. This phenomenon suggests a profound truth about the connection that extends into the fabric of quantum creation. Our intentions are not isolated; they exist within an interconnected energy field where our inner beliefs are reflected in our external world.

Entanglement serves as a mirror in manifestation. When we cultivate positive, intentional energy within, our environment echoes it back to us. Likewise, if we harbour doubt or fear, we often find ourselves mirrored in experiences that reinforce those feelings. Thus, we learn that to change our external reality, we must first change the energy within ourselves. The universe responds like a synchronised dance partner as we establish harmony within.

Walking through the bustling streets of Chennai, one can imagine how each person, each interaction, reflects the quantum entanglement of energies. The universe responds to our inner state, weaving experiences reflecting our subconscious beliefs. In this way, we are not just passive recipients of reality but active participants, shaping our surroundings through the energy we carry.

9. The Power of Observation: Selecting a Pathway Through Intentional Awareness

Returning to the double-slit experiment, where particles change behaviour under observation, we find a powerful lesson in quantum creation. Observation collapses potentialities, guiding a particle (or a desire) into a specific form. In our lives, focused awareness serves as this very catalyst. We can direct energy towards specific outcomes by consciously observing our thoughts, beliefs, and desires.

Manifestation is not merely a hope or a wish; it is the practice of intentional observation. When we focus on a goal, we energise it, giving it form and substance. It's as if, through our attention, we collapse the wave function of possibility, bringing a desired outcome into reality. This is the essence of co-creation: our conscious awareness acts as a beam of light, illuminating the path we wish to walk.

In these moments of quiet reflection, we recognise that every thought and intention we hold with awareness is a choice. It is our act of selecting from an array of possibilities, each waiting to come into being.

10. The Field of Infinite Possibilities: Tapping into the Quantum Field

At the heart of quantum creation lies the concept of the quantum field—a boundless field of energy where every possible outcome exists. As we have explored, this field embodies the principle of superposition, allowing all potential realities to coexist until one is chosen.

Manifestation invites us to align with this field and tap into its infinite possibilities. Visualisation, for instance, is one way we immerse ourselves in the quantum field. We connect with its frequency by vividly imagining our desires as real. In doing so, we signal to the field that this particular outcome will materialise in our experience.

Much like an artist who envisions a completed painting before setting brush to canvas, we are invited to hold our vision with clarity and intention. The quantum field responds to this clarity, gradually shaping the physical world to mirror the energy we project.

11. Resonance and Coherence: Harmonising with the Desired Frequency

Resonance and coherence are terms often used in physics to describe harmony and alignment within systems. In quantum creation, these principles align with our inner state and our external goals. When we experience coherence—when our thoughts, emotions, and beliefs are unified—we resonate at a frequency that naturally attracts similar energies.

Think of resonance as a tuning fork: when struck, it vibrates at a specific pitch, and nearby objects resonating at that frequency begin to vibrate in harmony. Similarly, our desires resonate within the quantum field, attracting energies that align with our intentions. By nurturing feelings of gratitude, joy, and love, we amplify this resonance, allowing us to magnetise experiences that reflect our inner state.

In quantum creation, the path to manifestation is less about force and more about harmony. As we cultivate coherence within, the universe aligns our outer world to match in its mysterious and infinite ways.

12. A Quantum Dance: The Art of Co-Creation

Quantum creation is not a solitary act but a dance with the universe. Every intention we set, every moment of presence, every act of kindness—each step in this journey aligns us more closely with the rhythm of creation itself. To manifest is to participate in this cosmic dance, to be both the dreamer and the dream, the observer and the observed.

In these reflections, we find a profound truth: we are not separate from the universe but intrinsically woven into its fabric. Every thought, desire, and breath is part of the quantum field, each a thread in the tapestry of existence. Manifestation is not an external process but an awakening to the creative potential within.

As the day comes to a close and the quiet fills the spaces between thoughts, there is a sense of awe for the boundless possibilities, the quantum field that pulses with life, and the inner power that lies waiting to be realised. To manifest is to remember that, in this intricate web of reality, we are creators, weaving our desires into the quantum field with each heartbeat, each whisper of intent.

And in that quiet moment, as one looks up at the stars or listens to the night sounds of the city, there is a knowing. A knowing that the universe has heard, that intentions are seeds already planted, and that, in time, with patience and trust, they will bloom in their own mysterious, beautiful way.

CHAPTER 8

The Quantum Embroidery: Weaving Neville Goddard's Concepts into the Fabric of Quantum Physics

In the quiet moments of reflection, as the mist lifts from the valleys and the first rays of dawn cast a golden hue over the hills, one can't help but feel the intricate web of connections that bind us to the universe. Much like the delicate dance of light and shadow that plays upon the mountain paths, the invisible thread of thought also connects us to the reality we create. In this chapter, we shall explore how some of Neville Goddard's profound insights intertwine with the mysterious world of quantum physics, offering us a glimpse into the power of the mind and the magic of manifestation.

Living in the End: The Quantum Observer Effect

Neville's 'Living in the End' concept beckons us to embrace the feeling that our desires have already come true. It's as if mentally living in a fulfilled state pulls the future into the present. Much like standing on a cliff's edge, staring into the horizon and seeing your future self-living the life you've always dreamt of, this idea is both exhilarating and empowering.

But How Does This Connect With Quantum Physics?

In the quantum realm, particles exist in a state of superposition, where all possible outcomes exist simultaneously until observed. This phenomenon is the

Observer Effect—when a conscious observer decides what reality will manifest by focusing on a specific result. If we are to view our desires as these particles, then living in the end becomes our act of observation. By choosing to feel and live as though we have already realised our dreams, we collapse the wave of potential futures into the reality we desire.

Example: Picture a man yearning for financial freedom. Instead of dwelling on his current circumstances, he chooses to live in the end—he visualises himself walking through a bustling city, carefree, knowing his bank account overflows. He feels the sun on his face and the satisfaction in his heart. In quantum terms, he's focusing on one probable outcome among the infinite, thus pulling that reality into his experience. Soon, opportunities, people, and 'coincidences' align, much like how particles fall into place when observed, bringing his vision to life.

The Bridge of Incidents: Quantum Entanglement in Action

Neville's 'Bridge of Incidents' speaks of the seemingly random series of events that unfold to lead us towards our desired outcome. Once we've set our intention, the universe responds, often in unexpected ways, akin to the quantum phenomenon of 'entanglement', where particles, though separated by vast distances, remain mysteriously connected, influencing one another in ways that defy classical physics.

You set a chain of events when you visualise and assume your desire is fulfilled. Like an orchestra conductor, the universe arranges a symphony of opportunities and connections, guiding you along a path you may not fully understand but will inevitably lead you to your goal.

For instance, imagine a woman who dreams of writing a book. She assumes her book is already published, feeling its weight in her hands and visualising the cover. Days later, she meets a stranger at a café who is an editor seeking fresh voices. The woman could have seen this encounter as pure luck, but the quantum entanglement at work is a bridge of incidents that the universe set into motion, guiding her to her dream.

States of Consciousness: Parallel Realities and Quantum Shifts

Neville's belief in 'States of Consciousness' reminds us that we constantly shift between different versions of ourselves, altering our external world. In quantum mechanics, we might relate this to the idea of 'parallel realities' or 'quantum states', where infinite versions of reality exist, but we experience the one that aligns with our current state of consciousness.

Every time you change your mindset or belief, you are, in essence, stepping into a different version of yourself in a parallel reality. Just as a quantum particle can occupy multiple states until observed, your consciousness oscillates between various realities until you choose one by your dominant state of mind.

For instance, a man who once believed he was unlucky in love always found himself in failed relationships. One day, he shifted consciousness, thinking he was worthy of deep love and connection. He began to see himself differently, and soon, his external reality mirrored this shift. Like a quantum leap, his life changed direction, and love blossomed where there once was meditation:

Prayer and Meditation: The Focused Energy of Quantum Coherence

For Neville, 'prayer' was not about pleading or begging but about entering a feeling where your desire has already been fulfilled. This concept dovetails beautifully with 'quantum coherence', where a focused, steady energy field allows particles to maintain alignment and act in harmony.

As Neville described, when we meditate or pray, we focus on a singular desire, creating coherence within ourselves. In the quantum world, this coherence allows particles to remain synchronised, preventing 'decoherence' (or scattering), which disrupts the flow of energy.

For example, a young woman, overwhelmed by financial burdens, prays not for relief but with the feeling that her debts have already been paid. She meditates on the peace and freedom of financial security. As her energy becomes coherent and focused, the quantum field responds, aligning events in her life to make this a reality—unexpected bonuses, opportunities, and solutions appear, clearing her debts with surprising ease.

The Sabbath: Non-Interference in Quantum Manifestation

The 'Sabbath' in Neville's teachings represents a state of rest, where you step back and allow the universe to do the rest after assuming the feeling of the wish fulfilled. This principle is much like the 'non-interference' required in quantum experiments. Once you have observed and collapsed a particular wave function into reality, further interference or worry can destabilise the outcome.

In practical terms, this means that after you've done the work of imagining and feeling your desire fulfilled, you must rest in the knowledge that it is done. Constantly checking or doubting the process is akin to interrupting the quantum field, which may cause delays or shift the outcome.

Let's say a man envisions himself with the job of his dreams; he lets go of the how and the when trusting that the proper series of events will unfold. He doesn't constantly check his email or second-guess his abilities. He rests in the knowledge that the job is his, much like how, in quantum emphysics, once a wave function collapses, further interference is unnecessary. Soon, the call comes, and his faith gets rewarded.

Mental Diet: Coherence Amidst Quantum Noise

Finally, Neville's 'Mental Diet' emphasises the importance of maintaining positive, aligned thoughts to ensure that your external world mirrors your desires. In the quantum realm, this is akin to preserving 'quantum coherence' amidst the noise. Just as particles in a coherent state can maintain a singular focus, our minds must guard against negative thoughts, which act like quantum noise, disrupting the flow of manifestation.

Imagine that a man constantly worries about losing his job. As he feeds these thoughts, his external world begins to reflect his fears—misunderstandings at work, missed opportunities, and strained relationships. When he changes his mental diet, focusing instead on gratitude and competence, the noise clears, and his reality shifts. His performance improves, and his career flourishes.

As we journey through life, much like a traveller making their way through the winding roads of the hills, we realise that the universe is both a mirror and a dance partner. Neville Goddard's teachings, when viewed through the lens of quantum physics, remind us that we are the observers, the creators, and the participants in this grand tapestry. By aligning our thoughts, feelings, and consciousness with the reality we wish to experience, we change our lives and tap into the quantum world's boundless potential.

As the mist lifts and the sun rises, remember—just as the hills and valleys shift in the changing light, your world also changes in response to the light of your thoughts.

Everything Is Connected, and Everything Is Possible

Like the journey of our lives, the road ahead is full of unexpected turns, gentle slopes, and occasional steep climbs. But in this quantum dance, as Neville Goddard so elegantly taught, we are not mere passengers. We are the creators of our experience, the weavers of our destiny. Our thoughts, emotions, and beliefs are the threads we use to knit the fabric of our existence, and just like a traveller with a clear destination in mind, we must trust in the path, even when the destination remains out of sight.

The Invisible Web of Thought: Quantum Entanglement in Our Daily Lives

Imagine for a moment that the universe is like a spider's web, intricately woven, delicate yet strong. Every thought and belief is a vibration on this web, sending ripples across the strands, affecting not just our reality but the realities of others, which is where Neville's teaching of 'Everyone is You Pushed Out' ties beautifully with the quantum idea of 'entanglement'.

Neville's concept suggests that the people we encounter and the situations we face are parts of our inner world, entangled with our beliefs, desires, and assumptions. Just as those quantum particles respond to one another across vast spaces, so too do our thoughts influence others, often in ways unseen.

For example, when you dreaded meeting someone, convinced they would be rude or dismissive. And lo and behold, when you meet them, they are as we expected to see them. In contrast, consider the moments when you've assumed the best of someone, believing they would be kind, helpful, or generous—and, as if by magic, they were. It's not magic but a profound interplay between your internal state and the external world, like the entangled particles mirroring each other's behaviour.

When you change your assumptions, you change how people respond to you. The next time you encounter a difficult person or situation, ask yourself, "What is this reflecting within me?" And just as you would tune an instrument to produce the right note, adjust your inner beliefs to create the external harmony you seek.

The Power of Assumption: The Quantum Leap into New Realities

Neville often spoke of the power of 'Assumption' — the idea that what we assume to be true in our inner world will eventually manifest in the outer world. Quantum physicists speak of a similar concept: the 'quantum leap'. In quantum mechanics, particles do not move gradually from one state to another; instead, they make sudden leaps between states. These leaps occur without any visible transition, much like how our lives can change dramatically, seemingly overnight, when we align our inner beliefs with our desires.

When we embody the feeling of our wish fulfilled, we are making a quantum leap in consciousness. This leap is often invisible to others – there are no visible steps or bridges, just a sudden shift in reality. This is why changes can seem instantaneous or unexpected, both to ourselves and those around us.

Here is an example - Picture a young man who has always struggled with his self-worth. He's had years of feeling invisible and unimportant. But one day, after listening to Neville's teachings, he changes his assumptions about himself. He begins to assume that he is valued and respected. He walks through his days with this assumption, not waiting for others to confirm it but simply embodying it. And then, one morning, he wakes up to a completely different reality. Job offers start pouring in, people treat him with newfound respect, and his relationships deepen. He's made a quantum leap—without any visible transition, his world has shifted to match his inner state.

There lies the beauty of the power of assumption. By changing your state internally, you collapse the infinite possibilities into the one you desire most, pulling it into your reality like a magnet.

Revision: Changing the Past, Influencing the Future

Perhaps one of Neville's most intriguing concepts is 'Revision' — the idea that we can go back and rewrite the past in our minds, and by doing so, we change our memories and future outcomes. In the quantum world, this is not so far-fetched. The **delayed-choice experiment** in quantum mechanics suggests that actions taken in the present can influence events that have already happened. It's as though time is not a straight line but a fluid, malleable force that bends to the observer's will.

In the same way, Neville taught that by revising past events—by imagining them as we wished they had happened—we change their impact on us in the present and, in turn, alter the course of our future.

Consider a woman who has always felt defined by a difficult childhood. She grew up in a household filled with criticism and neglect, and these experiences have shaped her into someone who feels undeserving of love and success. But then she learns of Neville's concept of revision. She revisits those painful memories each night before bed, but this time, she imagines them differently. Instead of hearing harsh words, she imagines her parents offering praise and encouragement. Instead of feeling neglected, she feels cherished.

Over time, this nightly revision practice changed her deeply held beliefs about herself. She starts to see herself as someone worthy of love and success. And as she changes her perception of the past, her future changes, too. Opportunities she never thought possible begin to present themselves, and she finds herself in a loving relationship, thriving in ways she had never imagined.

Quantum physics supports this idea with the understanding that the observer plays a crucial role in shaping reality. Just as the woman revised her memories, we too have the power to rewrite the past and, by doing so, change the trajectory of our future.

Feeling is the Secret: Emotions as the Key to Quantum Creation

At the heart of Neville's teachings is the simple yet profound truth: 'Feeling is the Secret.' It is not enough to merely think about what we want; we must 'feel' it as though it is already ours. In quantum physics, 'energy' is everything, and our emotions are powerful vibrations that can influence the quantum field. When we feel joy, gratitude, or love, we resonate with the high vibrations of creation, attracting experiences that match these feelings.

In the quantum world, like attracts like. Our feelings act as tuning forks, drawing to us the experiences that resonate with our emotional state. Neville emphasised feeling—it is the key to unlocking the power of the quantum field.

Please imagine a man who desires to travel the world. Instead of simply thinking about the places he wants to visit, he closes his eyes and feels the warm sand beneath his feet on a tropical beach. He imagines the scent of the ocean air, the sound of waves crashing, and the joy that fills his heart as he experiences his dream vacation. He does this every night before falling asleep, fully immersing himself in the feeling of his desire.

Soon, circumstances in his life begin to shift. A job bonus, an unexpected gift, and a friend with airline miles converge to make his dream trip a reality. He didn't just think about travelling—he felt it, and in doing so, he resonated with the energy of his desire, drawing it into his experience.

As we reflect on these profound intersections between Neville Goddard's teachings and the quantum world, it becomes clear that we are not passive spectators in this grand play of life. We are the creators, the dreamers, the architects of our reality. Like a skilled artist painting on the canvas of the cosmos, we use our thoughts, beliefs, and feelings to shape the world we experience.

In this quantum universe, where infinite possibilities await our observation, let us be mindful of the energy we project, the assumptions we hold, and the emotions we nurture. In this delicate dance between the seen and the unseen, between the mind and the quantum field, we can manifest a life beyond our wildest dreams.

And as the sun sets behind the hills, casting long shadows over the landscape, let us remember: the light of our thoughts, the power of our feelings, and the strength of our imagination can illuminate even the darkest paths, guiding us towards the reality we wish to create.

There's a peculiar magic in the air when you realise that life is more than what it seems on the surface. It's the kind of magic that comes to you in quiet moments—sitting by a window, watching the rain softly tap against the glass, or during the long walks where the world feels endless, and possibilities stretch beyond the horizon. In these moments, you wonder if reality is as fixed as told to us. Is life just a series of events happening to us, or could it be that we are shaping life, gently moulding it with our thoughts, feelings, and conversations?

Here is where Neville Goddard's teachings come to life, wrapped in the delicate folds of quantum physics. It's a delicate balance—a symphony—where imagination, faith, and inner conversations play the lead instruments. And like any good symphony, when these elements come together in harmony, the reality we've always dreamt of emerges as beautiful.

Imagination: The Blueprint of Reality

Neville's words often danced around one powerful idea: **Imagination creates reality**. At first, it might sound like a whimsical notion, the kind of thing you tell children when they dream of flying or living in castles. But when you look closer, you realise that this isn't just fantasy—it's a profound truth that quantum physics supports.

In the world of quantum mechanics, particles exist in a state of potential until we observe them. They are everywhere and nowhere at once, waiting for someone—or something—to give them direction. It's the observer who pays attention and collapses that potential into a concrete reality. And this is where our imagination steps in.

Take, for instance, a man who longs for success but feels trapped in a cycle of mediocrity. He goes through his days working at a small desk in a corner office, imagining that life will always look this way. But just before sleep takes him at night, he lets his imagination run wild. He sees himself standing before crowds, his words inspiring thousands. He imagines signing books with his name on the cover, feels the pen's weight in his hand, and even smells the ink. Night after night, he returns to this imagined scene. It becomes so natural to him that it feels like it's already happening. And then, in subtle and grand ways, opportunities begin to appear. Someone asks for his advice, and a publisher calls. Eventually, he lives the life he once only imagined.

In the quantum field, possibilities are endless, just as in our imagination. When we permit our minds to dwell on these possibilities, we collapse them into our reality, shaping life in a manner we may not fully understand but can certainly feel.

Faith: The Power of Believing in the Unseen

If imagination is the blueprint, then faith is the foundation. In Neville's world, 'faith in the unseen' was paramount. It wasn't enough to imagine; you had to believe, genuinely believe, that what you desired was already yours. Faith isn't unquestioning optimism – it's a deep knowing that, even if the physical world shows no evidence, what you've imagined is coming.

This concept finds a remarkable parallel in quantum physics. Just as particles exist in a state of potential, waiting for our observation, so do our desires in the unseen realms, waiting for our faith to bring them into focus. Faith is the observer, the force that collapses possibility into reality.

Consider a young woman who dreams of finding love. She spends years searching, going on dates that lead nowhere, and feeling disheartened. But then, one day, she decides to change her approach. Instead of seeking love, she begins to live as though it's already hers. She feels the warmth of a loving relationship in her heart, imagines the joy of shared moments, and trusts that love is on its way, even if she doesn't see it yet. Weeks turn into months, but she doesn't waver. And then, when she least expects it, love walks into her life as if drawn by the invisible thread of her faith.

In the quantum world, what we focus on grows. Faith is that unwavering focus. It's the quiet, persistent belief that, even when nothing seems to be happening, the universe is at work, rearranging itself to match our inner state.

CHAPTER 9

Integrating Science and Spirituality in Daily Life

The early morning mist clings to the hills, slowly giving way to the gentle warmth of the rising sun. In these quiet moments, where the boundaries between the tangible and the intangible blur, I find myself most connected to the essence of life. This connection forms the foundation of my daily routine, a blend of ancient wisdom and modern understanding that guides me through the dance of life. But this chapter is not just about my routine—it's about how you, the reader, can explore and integrate these concepts into your life, creating a unique path that aligns with your goals and aspirations.

Science and Spirituality Defined

To truly integrate science and spirituality into daily life, it's essential to understand the principles that underpin each.

Science, particularly quantum mechanics, offers us insights into the fundamental workings of the universe—concepts like the observer effect, superposition, and entanglement. These principles suggest that our thoughts and intentions can influence reality at a fundamental level. The observer effect, for example, indicates that the mere act of observing a situation can change its outcome, which has profound implications for how we think about creating our reality.

On the other hand, spirituality encompasses practices and beliefs that connect us to something greater than ourselves, whether through mindfulness, compassion, or the Law of Assumption. These spiritual principles guide us in aligning our inner world with the external realities we wish to create. For instance, the Law of Assumption, as taught by Neville Goddard, encourages us to assume the feeling of the wish fulfilled, thereby aligning our subconscious mind with our deepest desires.

These two realms—science and spirituality—provide a robust framework for manifestation and personal growth. While science gives us the mechanics of influencing reality, spirituality offers the tools and practices to do so ethically, compassionately, and in alignment with our highest good.

Living with Intention: The Foundation of Creation

When it comes to me, each day begins with breathwork, grounding me in the present-moment and setting the stage for intentional living. I close my eyes and practice pranayama—a controlled breathing technique that channels life-force energy through my body. On some days, I turn to the Wim Hof method, feeling the invigorating rush of oxygen infuse my cells, awakening every part of my being. This breathwork is more than just a physical exercise; it is the gateway to a state of focused intention.

With my breath steady and my mind clear, I move into meditation. This practice, rooted in ancient traditions, aligns me with the deeper currents of the universe. Here, in the stillness of my mind, I plant the seeds of my daily intentions. I visualise the day ahead with clarity, imagining each task and interaction unfolding gracefully and purposefully. This visualisation is not merely wishful thinking but a deliberate act of creation—a practice that aligns my inner world with the external realities I seek to manifest.

But this is my routine—one that suits my goals, rhythm, and lifestyle. For you, the reader, your journey might look very different. The tools and techniques I use, such as pranayama, meditation, and visualisation, are just a few among many that we've discussed in this book. Your path to integrating science and spirituality will be unique to you, shaped by your aspirations, challenges, and circumstances.

Exploring Quantum Tools: Find What Resonates with You

You might find that affirmations resonate deeply with you as a powerful method to reprogram your subconscious mind and align it with your goals. Or perhaps visualisation becomes your primary tool, helping you see and feel your desired outcomes as already achieved, thereby attracting them into your life. The key is to explore these quantum tools—affirmations, visualisation, meditation, intention-setting, and more—and discover which ones feel most natural and effective for you.

For some, mindfulness might be the cornerstone of their daily routine.

Mindfulness, or the art of being fully present, is a powerful way to integrate both life's scientific and spiritual aspects. It allows you to connect deeply with each moment, fostering a sense of peace and clarity.

Others may find that grounding themselves in the present moment through practices like deep breathing or mindful observation helps them stay centred and aligned with their intentions. For many, a combination of quantum tools like meditation, affirmation, self-love, inner child healing, and shadow work may be an ideal quantum manifestation method. In contrast, for others, it might just be visualisation and SATS. The choice of tools for quantum manifestation depends on the individual and the desired outcome.

Create Your Routine

Tailor Practices to Your Goals

Your daily routine should reflect your unique goals and aspirations. For example, Hal Elrod, in his book 'The Miracle Morning', created a routine called SAVERS—Silence, Affirmations, Visualisation, Exercise, Reading, and Scribing (journaling). This routine has become a powerful tool for many people, helping them start their day with clarity and purpose. You can also create a morning routine that suits your needs and enables you to move closer to your goals.

If your goal is financial abundance, you might start your day with affirmations focused on prosperity and abundance, followed by visualisation exercises where you see yourself achieving your financial goals. Suppose you dream of winning a medal in the Olympics. Your routine might involve intense physical training and mental practices like visualisation and intention-setting to keep you focused and motivated.

The Power of Presence: Mindfulness in Action

Incorporating mindfulness into your routine can be transformative as you move through your day. Mindfulness is not just about meditation; it's about being fully present in each moment, whether working, eating, or spending time with loved ones. This practice helps you stay connected to your intentions and prevents you from being overwhelmed by distractions or negative thoughts.

Mindfulness also fosters a deeper connection with the world around you. By being present, you can fully engage with your experiences, noticing the beauty in everyday moments and finding peace in the simplicity of being. This state of presence enhances your ability to align your thoughts, actions, and emotions with the reality you wish to create.

Incorporate gratitude into your morning or evening routine. You might keep a gratitude journal, listing what you are thankful for daily. Or, you could take a few moments each morning to mentally acknowledge the blessings in your life. This simple practice profoundly affects your mindset and ability to manifest your desires.

Core Practices for Integration

Living with Intention: The Foundation of Creation

Every day begins with intention, which means starting my day with breathwork, grounding myself in the present moment, and setting clear intentions for what I want to achieve. But your approach might be different. Living with purpose is about setting purposeful goals and aligning your actions with those goals. Whether through meditation, journaling, or simply taking a moment to reflect, the key is to start your day with clarity and focus.

For example, I use pranayama or Wim Hof breathing techniques each morning to invigorate my body and centre my mind. I then sit quietly in meditation, allowing the breath to guide me into a deep state of focus, where I plant the seeds of my intentions for the day. I visualise these intentions as already fulfilled, feeling successful and accomplished.

However, this is just my way. You might find a different practice that resonates more with you – perhaps writing down your goals each morning or simply setting your intentions before starting your day. The important thing is to align your energy with your desired outcomes from the very start.

The Power of 'As If': Embodying Your Desires

One powerful concept from quantum mechanics and spiritual teachings is 'living as if', which means embodying the feelings and actions of already having what you desire, when you act 'as if', you align your energy with the reality you wish to create, making it more likely to manifest.

For instance, if you aspire to be more confident in your work, you can begin by carrying yourself with the confidence you would have if you were already at the peak of your career. Speak, dress, and make decisions as if you are already that successful person, which doesn't just change how others perceive you—it changes how you perceive yourself, which shifts the energy you project into the world.

This practice helps bridge the gap between where you are and where you want to be. It's not about pretending; it's about aligning your current reality with the frequency of your desired outcome.

Embracing the Power of Letting Go: Trusting the Process

Letting go is a crucial yet often overlooked aspect of manifestation. When you release your attachment to specific outcomes, you allow the universe to work its magic without resistance, which doesn't mean giving up on your desires; it means trusting that what is intended for you will come in its own time.

In my journey, I've learned that there are times when it's necessary to step back and allow things to unfold naturally, which can be difficult, particularly when you get deeply invested in a particular outcome. But I've found that when I let go of the need to control every detail, the universe often delivers in even better ways than I could have planned.

For some, this practice might be the key to unlocking their full manifesting potential. It requires faith in the unseen forces at work and a deep belief that the universe always works in your favour, even when the results aren't immediately visible.

Living a Life of Gratitude: Cultivating Abundance

Gratitude is another essential practice that can be integrated into your daily life, regardless of your specific goals. Expressing gratitude shifts your focus from what you lack to what you already have, aligning you with the frequency of abundance. This practice improves your mood and well-being and attracts more positive experiences.

You might keep a gratitude journal, listing what you are thankful for daily. Alternatively, you could take a few moments each morning to mentally acknowledge the blessings in your life. This simple practice profoundly affects your mindset and ability to manifest your desires.

Each morning, I begin my day by listing things I'm grateful for, but you might incorporate gratitude into your life differently. Whether through a gratitude journal, mental acknowledgements, or simply expressing thanks throughout the day, this practice shifts my focus to the positive. It helps attract more of what I appreciate into my life.

Dr Joe Dispenza, a prominent figure in the intersection of neuroscience and quantum physics, often speaks about the powerful effects of gratitude on the brain. According to his research, when we experience gratitude, the brain releases neurotransmitters like dopamine and serotonin, enhancing our mood and overall well-being. On a quantum level, this shift in focus and feeling alters the wave patterns we emit into the quantum field, attracting more of what we are grateful for.

Incorporating gratitude into your daily life can be as simple as writing down three things you are grateful for each morning or evening. By focusing on specific instances—such as the support of a loved one, the beauty of a sunrise, or a minor success at work—you deepen the emotional impact of gratitude, thereby amplifying its vibrational power.

The Role of Meditation: Bridging Consciousness and Subconsciousness

Meditation is a bridge between your conscious desires and your subconscious beliefs. In my routine, meditation helps me align with the deeper currents of the universe. For you, it might be a tool for clarity, relaxation, or spiritual connection. Regardless of how you practice it, meditation is a powerful way to align your inner state with your external goals.

In my daily practice, I use meditation to quiet the mind and focus on my intentions. By doing so, I create a space where my subconscious mind can absorb these intentions without interference from the doubts and fears that often plague our conscious thoughts, which allows me to move through the day with greater clarity and purpose.

The key is consistency. Whether you meditate for five minutes or an hour, regular meditation helps you stay connected to your inner self and aligned with your goals, making it an essential part of integrating science and spirituality into daily life.

The Science of Emotions: Resonating with Reality

Emotions are not just fleeting feelings but energy fields that influence our reality. Practices that cultivate positive emotions—

joy, love, and gratitude—raise your vibrational frequency, aligning you with the experiences you want to attract. Understanding the science behind emotions can help you harness their power to create your desired life.

In my experience, emotions like joy and gratitude have the most significant impact on my reality. When I focus on these emotions, I notice that my day flows more smoothly, opportunities present themselves more readily, and I feel more connected to the people around me. Cultivating these emotions can be as simple as taking time each day to do something you love, spending time with loved ones, or practising gratitude.

By understanding that emotions are powerful tools for manifestation, you can consciously choose to cultivate the emotions that align with your desires, which enhances your well-being and amplifies the effectiveness of your manifesting practices.

Living a Life of Purpose: Aligning with Universal Laws
The purpose is the driving force behind true fulfilment. When you live in alignment with your higher purpose, your actions resonate with the universal laws of creation, making it easier to manifest your desires. Whether your purpose is to serve others, achieve personal growth, or create something new, aligning with it brings clarity and direction to your life.

In my life, I've found that when my actions align with my purpose, the universe supports me in ways I could never have imagined. This alignment brings a sense of fulfilment beyond material success – it's about knowing that I'm contributing to something greater than myself.

Your purpose might be different from mine, and that's perfectly okay. What's essential is that you find what drives you—whether it's contributing to your community, achieving personal growth, or pursuing a career that fulfils you—and live in alignment with that perfectly.

Ethical and Holistic Approaches

Aligning with Universal Harmony: Creating for the Greater Good
When manifesting your desires, it's essential to consider how they align with the greater good.

Ensuring that your goals contribute positively to the collective consciousness creates a better life for yourself and those around you. This approach fosters a sense of interconnectedness and helps maintain balance in the universe.

Imagine your desires as ripples in a pond. When you set an intention that aligns with universal harmony, those ripples spread outward, positively impacting the lives of others. It is not about sacrificing your needs but integrating them into a larger purpose. For instance, if your goal is financial abundance, consider how this abundance can benefit others—whether through charitable contributions, creating jobs, or supporting causes that matter to you.

The Power of Compassion: A Guiding Principle

Compassion is a critical element of ethical manifestation. When empathy and kindness guide your actions, you create positive energy that ripples into the world.

Compassion ensures that your manifestations are not just self-serving but contribute to the well-being of others.

This principle is fundamental when considering all things' interconnectedness. By acting compassionately, you align yourself with a higher frequency of love and understanding, attracting more positive experiences. Compassionate action might involve helping a friend in need.

- Volunteering in your community.
- Simply being kind to yourself during challenging times.

The Responsibility of Co-Creation: Nurturing a Positive Inner World

As a conscious creator, you are responsible for cultivating a positive inner world. Like a gardener tends to their garden, you must nurture thoughts, emotions, and beliefs supporting your highest good. This responsibility extends to how your energy affects others, emphasising the importance of maintaining a healthy, loving, and joyful inner environment.

The energy you put into the world reflects your inner state. When you focus on cultivating positivity within, you naturally attract positive experiences, which doesn't mean you won't face challenges. Still, it means you'll get better equipped to handle them gracefully and resiliently. By maintaining a positive inner world, you also contribute to the collective well-being, raising the vibration of those around you.

Journey of Self-Discovery: Deepening Awareness

Integrating science and spirituality is not just about achieving goals; it's also a journey of self-discovery. As you explore these practices, you'll gain deeper insights into who you are and what drives you. This process of self-awareness and growth is essential for living a life that is not only successful but also meaningful.

Self-discovery involves asking the big questions: Who am I? What is my purpose? How can I contribute to the world? These questions may not have immediate answers, but exploring them will lead to greater self-awareness and a deeper connection to your true self. As you integrate science and spirituality into your daily life, you'll find that your journey of self-discovery becomes more prosperous and more fulfilling.

Action and Integration

Taking Inspired Action: Moving Towards Your Goals

Manifestation is not a passive process. It requires inspired action—taking steps that align with your intentions and moving forward with purpose and confidence. Whether starting a new project, learning something new, or simply taking a small step towards your goal, action turns your intentions into reality.

Taking inspired action means committing to my daily routines, pursuing new opportunities that align with my goals, and continuously learning and growing. For you, inspired action might look different. It could be reaching out to a mentor, starting a new habit, or taking a leap of faith in a new direction. The key is to act with intention, trusting that each step brings you closer to your desired outcome.

Embracing Synchronicity: Following the Signs

As you take action, be open to the synchronicities that appear along the way. These are the universe's way of guiding you towards your goals, often through seemingly random coincidences or unexpected opportunities. Pay attention to these signs, trust your intuition, and follow the path that unfolds before you.

Synchronicities are more than just coincidences – they are meaningful alignments that reflect your inner state. For example, you might think about a friend you haven't spoken to in years, and then they suddenly call you. Or you might be seeking a solution to a problem, and the answer appears in the form of an article, a conversation, or a chance encounter. By remaining open to these signs, you allow the universe to guide you towards your goals in ways you might not have expected.

Celebrating Milestones, Big and Small: Maintaining Momentum

The manifestation journey can be long and challenging, so it's important to celebrate your progress along the way. Acknowledging and celebrating milestones—no matter how small—keeps your motivation high and reinforces your belief in the process. These small victories keep you moving forward, even when the road gets tough.

In my life, I've found that taking the time to celebrate achievements—whether completing a challenging project or simply maintaining a new habit—helps maintain momentum and motivation. These celebrations don't have to be grand; even a tiny acknowledgement of progress can make a big difference in your journey. Celebrating your milestones reinforces the positive energy that propels you toward your goals.

Living a Life of Integrity: Building a Strong Foundation

Integrity is the foundation of all successful manifestations. When you align with your values and stay true to your word, you create a robust and stable foundation for manifesting your desires. Integrity ensures that your actions are consistent with your intentions, creating a positive ripple effect that extends far beyond your personal life.

Living with integrity means making choices that align with your highest self, even when challenging. It involves being honest with yourself and others, following through on your commitments, and acting in ways that reflect your values. When you live with integrity, you strengthen your manifesting power and inspire others to do the same.

Conclusion: Living a Quantum Life

As I sit here in the fading light of the evening, the Chennai skyline glowing softly in the distance, I am filled with a sense of peace and purpose. The journey of integrating science and spirituality is not a destination but a continuous dance – a dynamic interplay between the mind and the heart, the known and the unknown.

Living a quantum life means embracing this dance, recognising that each day offers new opportunities to align with the creative forces of the universe. It means starting each morning with purpose, moving through the day with presence, and ending each evening with reflection and gratitude. This path is deeply personal and ongoing, and it offers a way to navigate the complexities of life with grace, wisdom, and an open heart.

May your journey be filled with wonder and discovery, with moments of profound insight and quiet reflection. Remember, you are not just a witness to the universe but a co-creator, shaping your reality with every thought, every intention, and every action. Live with purpose, act with compassion, and trust in the infinite possibilities that await you. The journey is yours to create, and the universe is ready to support you in ways you have yet to imagine.

CHAPTER 10

Embracing the Quantum Self: Unity, Consciousness, and the Interconnected Universe

In the hidden depths of reality lies a world we're only beginning to grasp—a world governed not by division and separation but by unity, potential, and infinite possibility. The quantum world offers us more than a glimpse into the mysteries of particles and energy; it offers us a transformative understanding of consciousness and existence. It's a journey away from the old mechanical model, where the universe was a grand machine ticking on relentlessly, and toward a more holistic, connected reality that honours the dynamic web binding us all together.

Breaking Away from the Mechanical View

For centuries, science followed a model of separateness, the 'mechanistic' worldview. In this view, the universe operated like a giant clockwork mechanism. Everything comprised separate, static pieces, each interacting only through force. Change, in this model, required an external push, a shove, to bring about movement and life. This was the world of classical physics, where Newton's laws governed everything, from the fall of an apple to the paths of planets.

But as physicists explored deeper, they encountered strange phenomena that didn't fit. These new insights sparked a revolution, shattering the foundations of mechanistic thinking. Quantum physics revealed the world as a dance of possibilities rather than rigid certainties. It showed that, at its most fundamental level, reality is woven from probabilities, potential, and interconnectedness.

The Quantum Leap: Seeing Potential Instead of Static Objects

In quantum mechanics, particles don't exist in fixed states; they exist as potentialities until observed. This concept, known as superposition, means that an electron, for instance, isn't 'here' or 'there' until observed—it exists in multiple states simultaneously. When we observe, we aren't merely passive onlookers but active participants shaping reality.

This revelation challenges the idea of a world made of 'things'. Instead, the universe is better seen as an ocean of possibilities, with every particle existing as a probability wave until it 'collapses' into a defined state. It's a dance of becoming rather than being, of potential rather than permanence. And in this dance, consciousness itself plays an active role.

Consciousness as the Foundation

Max Planck, often considered the father of quantum theory, once remarked, "I regard consciousness as fundamental. I regard matter as derivative from consciousness." This notion flips traditional thinking on its head. Rather than seeing consciousness as a by-product of physical processes, it suggests that consciousness is the underlying reality from which all else emerges. In this view, matter and energy are expressions of consciousness.

Planck's insight redefines our place in the cosmos. If consciousness is primary, then we, as conscious beings, are not separate entities in a vast, indifferent universe. Instead, we are part of the fabric of reality—consciousness exploring and experiencing itself. This view aligns beautifully with ancient wisdom traditions, which teach that we are not isolated but interconnected expressions of a universal consciousness.

Entanglement: The Universe as an Indivisible Whole

One of the most profound discoveries in quantum mechanics is the concept of entanglement. When two particles become entangled, they remain connected no matter how far apart they are, behaving like a single entity. This

Defies classical notions of separateness and distance. Imagine two particles on opposite sides of the galaxy responding to each other instantaneously; their states remain intertwined, bound beyond the limits of space and time.

This phenomenon led physicist David Bohm to propose that separateness is an illusion. Bohm viewed the universe as a single, undivided whole, with everything interconnected fundamentally. He compared it to a hologram, where each part contains the whole. He argued that the reality we perceive as fragmented and individualistic is merely a surface view, concealing a more profound, unified reality.

In Bohm's vision, particles are not independent 'things' but manifestations of a more profound 'implicate order'. Like a vast web of connections, this hidden order is the reality from which the visible universe unfolds. Here, separateness dissolves, and we begin to see ourselves as integral parts of a greater whole, like drops in an ocean—each distinct yet inseparably linked to the vastness.

The Role of the Observer: Participation Over Passivity

The quantum realm reveals another startling truth: the observer matters. In classical physics, observation is passive; in quantum mechanics, it's transformative. When we observe a quantum system, we influence its behaviour, collapsing its wave function into a definite state. This phenomenon, often described as the 'observer effect', implies that consciousness interacts with reality at a fundamental level.

This dynamic interaction challenges the notion of an objective reality that exists independently of us. Instead, it suggests a participatory universe where our thoughts, intentions, and consciousness shape the fabric of reality. The boundary between observer and observed dissolves, inviting us to recognise our active role in unfolding the universe. We are not mere spectators; we are co-creators.

This participatory model aligns with ancient philosophies that teach the unity of self and universe, echoing the words of the Vedas: "Tat tvam asi"—"Thou art that." Here, the universe is not a cold, separate entity but a living, breathing expression of consciousness, with each of us contributing to its evolution.

Non-Locality: Transcending Space and Time

One of the most intriguing aspects of quantum mechanics is non-locality, the idea that particles can affect each other instantaneously across vast distances. Albert Einstein famously called this 'spooky action at a distance', as it appeared to violate the principles of causality and the speed of light.

Non-locality suggests that space and time are not the fundamental structures of reality but rather emergent properties of a deeper, interconnected field. This interconnectedness is not bound by distance limitations, indicating a level of reality where everything exists as one unified whole.

This insight has profound implications for our understanding of consciousness and interconnectedness. If separateness is an illusion, we are all connected at a fundamental level, transcending the physical boundaries we perceive. Our thoughts, emotions, and actions ripple through this unified field, influencing and interacting with the world in ways we are only beginning to understand.

Embracing Uncertainty: The Beauty of Indeterminism

The quantum world is governed by uncertainty, as articulated in Heisenberg's uncertainty principle. This principle states that specific properties, like a particle's position and momentum, cannot be precisely known simultaneously. This indeterminism challenges the predictability of classical physics, embracing a universe where uncertainty is fundamental.

Rather than viewing uncertainty as a limitation, quantum mechanics invites us to see it as a realm of possibility. The absence of fixed outcomes creates a space for creativity, spontaneity, and growth. Life is not predetermined in this space but a continuous unfolding of potential, allowing for new paths, discoveries, and transformations.

This openness to the unknown mirrors the path of personal growth and self-discovery. Embracing uncertainty encourages us to step beyond the familiar, trusting in the process of life itself. It reminds us that fixed identities or predetermined paths do not define us but are dynamic beings capable of endless transformation.

Wholeness Over Reductionism

In classical science, understanding a system meant breaking it down into its constituent parts – a process known as reductionism. Quantum mechanics reveals that wholes cannot always be understood by examining individual parts. Instead, the relationships between parts give rise to new properties and behaviours, emphasising the importance of interconnectedness.

Werner Heisenberg, one of the pioneers of quantum mechanics, described the universe as "a complicated tissue of events." In this view, everything influences everything else, creating a tapestry of connections that cannot be reduced to isolated elements. This holistic perspective aligns with the concept of "wholeness and the implicate order," as Bohm termed it, where the universe's unity shapes its parts.

In this paradigm, we are not isolated beings but integral threads in the fabric of existence. Our actions, thoughts, and energies ripple through the cosmos, connecting us to the whole. Embracing this wholeness shifts our perspective from individual gain to collective harmony, inviting us to live with compassion, empathy, and awareness of our interconnectedness.

The Seed of Potential: Living in Harmony with Nature

Nature exemplifies quantum principles in ways that science is only beginning to understand. Take the humble seed, which holds the potential for a towering tree within it. The seed does not struggle to become; it simply unfolds, expressing its innate potential in harmony with its environment. This process mirrors the quantum world, where interactions and relationships make a potential reality.

In this light, we see that growth and evolution are not about force or control but about unfolding and nurturing potential. This perspective encourages us to live in harmony with nature, recognising that we are not separate from it but integral parts of its self-organising intelligence. Just as an implicate order guides the seed's journey, so is our life path shaped by deeper currents of consciousness.

A Universe of Conscious Creation

Quantum mechanics teaches that the universe is a field of possibility shaped by consciousness, interconnectedness, and continuous transformation. As we expand our awareness of this quantum reality, we see ourselves not as isolated entities but as conscious creators in a shared cosmic dance. We participate in an unfolding story where each thought, action, and intention contributes to the whole.

In embracing this view, we step into a space of conscious creation. We recognise that our lives, like particles in the quantum field, are not fixed but filled with potential waiting to be realised. By aligning with this potential, we awaken to our true nature as expressions of universal consciousness, capable of manifesting-dreams, transforming challenges, and co-creating a world that reflects our highest aspirations.

As we journey into the mysteries of quantum consciousness, we are invited to rewrite the story of our lives. Each of us is a co-creator, a spark of consciousness, and a participant in reality unfolding. The quantum world reveals that we are more than mere observers; we are connected to a field of potential, shaping reality through intention and awareness.

This journey asks us to release outdated narratives of separation and control. As we align with the universe's interconnected nature, we find ourselves part of an endless symphony, echoing the wisdom of ancient traditions that recognise unity, compassion, and creativity as our true essence. Each thought and action ripples outward, reminding us of our immense power as conscious beings.

In honouring the principles of the quantum world—unity, potential, and transformation—we embark on a path of conscious creation. We become active participants, embracing the uncertainties and possibilities of life, knowing that within every moment lies the potential for growth, discovery, and transformation. This is the essence of quantum creation: an invitation to live with awareness, to shape reality with love, and to contribute to a world rooted in harmony and collective evolution.

Through this journey, we discover that the true essence of the quantum self is not mastery over the universe but a harmonious relationship with it, a compassionate engagement that reflects the interconnected nature of all existence. In embracing this wisdom, we step into our power as co-creators, cultivating a world that honours the mystery and beauty of consciousness and the infinite possibilities waiting to unfold.

This is the call of the quantum field: to see ourselves in the universe and the universe within us, each a vital part of the infinite whole. The journey begins with a single, conscious choice—to awaken to our potential and live in alignment with the interconnected, wondrous reality of quantum creation.

CHAPTER 11

The Mirror of Life: Everyone is You, Pushed Out

The world, with all its people and experiences, reflects the self. This simple, quietly profound truth has constantly stirred me—like the way a mountain stream, clear and still, reflects the trees and the sky above. The idea that 'Everyone is You Pushed Out' is akin to standing before the stream, watching the mirrored world, only to realise your inner landscape gets reflected at you.

Neville Goddard, a wise soul, once said that our assumptions shape life and nothing in this world is separate from us. All is connected, just as we can tie the sun to the shadows it casts. We are, at once, the creators and the participants of our reality. And the people we meet? Why are they not mere strangers or happenstance encounters? They are our thoughts and beliefs made manifest.

Let us wander together through this idea, much like a quiet walk through the hills. Imagine for a moment, as you stroll beneath the dappled sunlight filtering through the trees, that every person you meet is a reflection of you. It was a strange thought at first, but it stayed with me.

You see, we live in a world of mirrors. Not the cold glass kind, but mirrors woven from the threads of our inner selves. When you believe something—about yourself and the world—this belief quietly steps outside you and becomes the world you see. It is as if the trees lean a little closer, and the wind whispers a little softer to remind you that what you hold within shapes what lies without.

The Gentle Observer: A Quantum Parallel

Now, let's drift into another landscape of science and mystery. In quantum physics, there is a curious thing called the observer effect. It tells us that observing a tiny particle, for instance, can change its behaviour. Looking and expecting to bend reality to meet the observer's gaze.

Isn't that something? A simple glance, a fleeting thought, could ripple through the unseen world and make it dance to your tune. And so it is with life. The way you see the world, the way you believe it to be, changes everything around you.

This is where quantum physics meets the wisdom of Neville Goddard. 'Everyone is You Pushed Out' is much like this observer effect: what you hold in your heart and mind becomes the essence of the world you experience.

If you hold kindness within, the world will offer you kindness. If your heart gets burdened with fear or doubt, you will find these shadows in the faces and words of others. In some strange, beautiful way, we are both the painter and the canvas. Our assumptions, like brushstrokes, colour the world around us, and we paint those we meet with our deepest beliefs.

The Quiet Revolution Within

A quiet revolution takes place when you understand this. It is not a thunderous change but more like the first light of dawn creeping over the horizon—slow, gentle, yet powerful. Everything changes when you see that life is not happening to you but through you.

You might notice that the people in your life—the ones you hold close and even the fleeting encounters—begin to behave differently. The difficult colleague may soften, the distant friend may suddenly grow closer, and even strangers seem to carry a warmth that wasn't there before. It's not that they have changed. It's what you have.

The moment you change your internal landscape, the external world follows suit, much like the clouds parting to reveal the sun. And this change begins with a simple shift in belief. If you believe you are unworthy, the world will reflect this back to you. But if, instead, you begin to hold within yourself a quiet certainty of your worth, you will find the world reflecting this new belief, like a mirror catching the glow of the setting sun.

The Hills We Climb in Our Relationships

I've realised that relationships are the most precise mirrors of all. They are like the winding paths up the mountains—sometimes steep, sometimes gentle, always revealing something new about ourselves. If you repeatedly find yourself in the same kind of relationship, where your needs are unmet, or conflict arises like dark clouds before a storm, it may be time to look within.

It is not that the world has conspired against you, but rather that your inner beliefs get revealed through these encounters. If you believe, deep down, that you are not worthy of love, you will find yourself entangled in relationships where love is hard to come by. But here's the good news: you can change this, not by forcing the world to be different but by changing how you see yourself.

Once you believe in your worth, the world mirrors this belief. Those relationships that once felt like storms on the horizon now become peaceful, even beautiful. It's as if the mountains have shifted, making the path smoother underfoot.

The Journey of Assumption: A New Path

Just as the observer in quantum physics changes reality by simply being there, you, too, can change your life by shifting your assumptions. The fact you experience is one of many possibilities. It is like standing at a crossroads in the forest, with countless paths stretching out before you. Your assumptions determine the path you carry with you.

If you assume the world is a harsh place filled with difficulty, that is the path you will walk. But if you believe life is a place of beauty and possibility, that path will open up before you, winding gently through the trees, with sunlight filtering through the leaves.

The essence of 'Everyone is You Pushed Out' means that you do not need to force change upon the world. Instead, you gently shift your inner beliefs, and the world will follow. It is as if the very fabric of reality bends to meet your expectations. The hills rise and fall in response to your steps, the trees lean in to hear your thoughts, and the people you meet reflect the story you are telling yourself.

A Quiet Reflection: The Power of Belief

As we come to the end of this quiet walk, take a moment to reflect on the mirrors in your own life. What do they show you? Are you content with the reflection, or does it ask for something more? Remember, you hold the brush, and the world is your canvas. The beliefs you carry shape your reality's colours and strokes.

And as you walk through life, remember that every person you meet is part of that painting, reflecting the beauty of the struggles within you. The hills may be steep at times, but you can change the landscape with each step.

So, as you gaze into the mirror of life, know this: you are both the creator and the observer, and the world will always quietly reflect the story you choose to tell.

PART 2

Tools, Applications, and Mastery of Quantum Creation

CHAPTER 12

The Quantum Toolbox – Crafting Your Reality

"The only limits to the possibilities in your life tomorrow are the 'buts' you use today."

– Les Brown

Neo stood on the threshold of reality, his eyes reflecting quiet turmoil. Before him, Morpheus extended two small pills—one blue, one red—each glistening faintly under the muted light, like secret keys to realities unknown. The red pill seemed to pulse with truth so potent it threatened to unravel everything Neo had ever known. It was a truth that, once grasped, he could never let go. There he was, suspended in that singular moment of indecision that we all recognise—when we stand at the crossroads, teetering between the safety of the familiar and the beckoning realm of infinite possibilities.

In the movie *The Matrix*, Neo's journey from the mundane into a world where the mind holds the power to shape reality is more than just a science fiction fantasy. It's a reflection of our potential, our power to transform our lives. And here, in this chapter, we invite you to take your red pill—not a literal one, but a leap into the world of quantum creation. In this world, your thoughts, beliefs, and emotions are your tools, your instruments, waiting to be wielded with intention and care to craft the life you desire.

The Quantum Field: A Canvas of Infinite Possibilities

Imagine standing at the edge of a vast ocean that glimmers under a radiant sun. The surface is calm, yet infinite waves ripple beneath it, each representing a distinct possibility and reality waiting to unfold. This ocean is the quantum field—a boundless expanse of potential where your thoughts, beliefs, and emotions cast ripples that can shape your reality. In this field, nothing is fixed; everything is fluid, waiting for you to take the helm and steer your life towards the shores of your desires.

The Power of Visualisation – Painting Your Desired Reality

"The mind is everything. What you think, you become."

– Buddha

Visualisation is more than just daydreaming; it's a deliberate and powerful technique that creates vivid, detailed images of the life you want to lead. By mentally rehearsing the achievement of your goals, you engage your mind in a process that can lead to real-world outcomes. The science behind visualisation lies in its ability to rewire your brain, making your desires feel more attainable and guiding you towards action.

What Is Visualisation?

Visualisation creates a mental image of a future event, goal, or desired outcome. This practice taps into the brain's ability to simulate experiences, allowing you to practice and refine your actions before they occur. Visualisation isn't just for athletes or performers—it's a tool anyone can use to enhance their life.

When you visualise, your brain engages similarly to how it would if you were performing the activity. This phenomenon is why visualisation can be so powerful; it conditions your mind and body to succeed before you even take the first step. The more you visualise with detail and emotion, the stronger the mental connections you create, making the desired outcome more likely to manifest in your life.

How to Visualise

Visualisation is a practice that one can master with consistency and dedication. Here's a step-by-step guide to help you get started:

1. **Find a Quiet Space:** Choose a calm and peaceful environment where you won't be disturbed, such as a quiet room, a park, or any place where you feel relaxed.

2. **Relax Your Mind and Body:** Close your eyes and take deep breaths. Allow your body to relax and your mind to become calm. Let go of distractions and focus solely on your visualisation.

3. **Create a Clear Picture:** Create a mental image of your goal or desired outcome. Be as detailed as possible. If you visualise success in your career, see yourself in the role you desire, feel the satisfaction of achieving your goals, and imagine the positive reactions from those around you.

4. **Engage All Senses:** Visualisation is most effective when you engage all your senses. Don't just see your success—hear the sounds, feel the textures, and even smell the environment of your desired outcome. The more senses you involve, the more accurate the visualisation becomes.

5. **Feel the Emotions:** Connect emotionally with your visualisation. Feel the joy, pride, or relief of achieving your goal. These emotions amplify the impact of your visualisation, making it more likely to influence your reality.

6. **Practice Daily:** Consistency is vital. Dedicate time each day to your visualisation practice. Over time, it will become easier and more effective.

Michael Phelps, the most decorated Olympian with 28 medals, didn't rely on physical training alone to reach his remarkable achievements. Visualisation played a crucial role in his success. Phelps is renowned for visualising every aspect of his races, from stepping onto the starting block to the final touch at the wall. He would mentally rehearse his performance, imagining every stroke, turn, and finish with precision.

Phelps didn't just see the perfect race—he also visualised challenges, like his goggles filling with water, and mentally practised overcoming them. This preparation allowed him to remain calm and focused during actual races, no matter what obstacles arose. His visualisation practice became so ingrained that when his goggles filled with water during the 2008 Beijing Olympics, he could swim the rest of the race unthinkingly and still win gold, setting a world record.

Phelps's story is a powerful testament to the impact of visualisation. By consistently imagining success, he trained his mind and body to perform flawlessly, even under extreme pressure. Visualisation was a key component of his strategy, contributing to his extraordinary Olympic legacy.

Neuroscience research supports the effectiveness of visualisation. Studies have shown that when you visualise an action, your brain activates the same neural pathways as if you were performing the activity. This mental rehearsal strengthens these pathways, making the real-life execution of the task smoother and more natural. Essentially, by visualising success, you are training your brain to succeed.

Exercise: Painting Your Dreamscape

1. **Set the Scene**: Find a quiet, comfortable place to close your eyes and focus.
2. **Centre Yourself**: Take deep breaths to relax your body and calm your mind.
3. **Visualise Your Goal**: Create a vivid image of your desired outcome. Make it detailed and realistic.
4. **Engage Emotionally**: Feel the emotions that would accompany the achievement of your goal.
5. **Repeat Daily**: Dedicate at least 10 minutes daily to this practice, strengthening your visualisation skills and bringing your goals closer to reality.

Affirmations

"It's the repetition of affirmations that leads to belief. And once that belief becomes a deep conviction, things happen."

– Muhammad Ali

Affirmations are more than just positive statements; they are potent declarations that shape your reality. By consistently repeating affirmations, you align your thoughts, emotions, and actions with your desired outcomes. These statements are not mere words—they are tools that can reprogram your subconscious mind, fostering a mindset that supports your goals and dreams.

What are Affirmations?

Affirmations are positive statements you repeat to yourself, intending to bring about a desired change in your life. They reinforce specific beliefs and behaviours, effectively rewiring your brain. The more you repeat these affirmations with conviction, the more they become ingrained in your subconscious mind, influencing your thoughts, emotions, and actions.

The key to successful affirmations is consistency and emotion. It's not enough to say the words; you must believe in them and feel the feelings associated with the outcome you desire. When practised regularly, affirmations can shift your mindset, helping you overcome limiting beliefs and paving the way for personal and professional growth.

How to Practise Affirmations

Practising affirmations is a simple yet profound process. Here's a step-by-step guide to help you harness the power of affirmations in your daily life:

1. **Choose Your Affirmation**: Start by identifying a specific area of your life that you want to improve. Your affirmation should be positive, present tense, and exact. For example, instead of saying, "I will be successful," say, "I am successful."

2. **Engage Emotionally**: As you repeat your affirmation, connect with the emotions that accompany the outcome you desire. Feel the joy, pride, or satisfaction as if your affirmation is already true.

3. **Repeat Consistently**: Consistency is crucial. Set aside time each day to repeat your affirmation—ideally in the morning or before bed. Repetition strengthens the neural pathways associated with your affirmation, making it more effective.

4. **Use Visualisation**: Combine your affirmations with visualisation. As you repeat your affirmation, visualise the scenario where it's true. This combination reinforces the message in your subconscious mind.

5. **Believe in Your Affirmation**: Even if it feels untrue initially, continue to repeat your affirmation with conviction. Over time, your mind will accept it as reality, and your actions will align with this new belief.

Muhammad Ali, widely regarded as one of the greatest boxers of all time, didn't just achieve his status through physical prowess alone—his success had deep roots in the power of affirmations. Ali famously declared, "I am the greatest," long before he had ever won a championship. This wasn't just a boast but a deliberate and powerful affirmation he repeated to himself and the world.

Ali's affirmations were more than words; they reflected his unshakeable belief in greatness. By consistently affirming his identity as the greatest, Ali convinced himself and those around him. His affirmations shaped his mindset, giving him the confidence and determination to pursue his goals relentlessly.

Ali's affirmations kept him focused and resilient despite challenges and setbacks. He knew that declaring his greatness was forging his reality with every word. This practice was so effective that it became a crucial part of his mental preparation for every fight. Ali's legacy is a testament to the power of affirmations—he became the greatest through his words.

The science of neuroplasticity supports the power of affirmations. Neuroplasticity is the brain's ability to reorganise itself by forming new neural connections throughout life. When you repeat affirmations, you engage in a process that strengthens the neural pathways associated with positive beliefs and behaviours. Over time, this brain rewiring can change your thoughts, emotions, and actions, making it easier to achieve your goals.

Exercise: Crafting and Repeating Your Affirmations

1. **Identify Your Goal**: Think about an area of your life where you want to see improvement or change. This area may relate to your career, relationships, health, or personal growth.

2. **Create a Positive Affirmation**: Write a positive, present-tense statement that reflects your desired outcome. For example, if you want to improve your self-confidence, your affirmation could be, "I am confident and capable."

3. **Engage Emotionally**: As you repeat your affirmation, connect with the feelings of having already achieved your goal. Feel the confidence, happiness, or peace that comes with it.

4. **Repeat Daily**: Dedicate at least five minutes daily to repeating your affirmation. You can do this in front of a mirror, during meditation, or as part of your morning routine.

5. **Visualise Your Success**: While repeating your affirmation, visualise yourself living the reality of your statement. Imagine every detail, and let the visualisation reinforce your belief.

The Power of Journaling and Scripting – Writing Your Reality into Existence

"Writing is the painting of the voice."

– Voltaire

Journaling and scripting are powerful tools that allow you to externalise your thoughts, feelings, and desires, translating them from the intangible world of the mind into the physical world of words.

These practices serve as a bridge between your inner world and the reality you wish to create. By putting your thoughts into writing, you clarify your intentions and set the stage for their manifestation.

What are Journaling and Scripting?

Journaling involves regularly writing down your thoughts, experiences, and emotions. It is a reflective tool that helps you gain insight into your life, recognise patterns, and make sense of your journey. Journaling allows you to process emotions, set goals, and track progress.

Scripting, however, is a specific form of journaling that involves writing about your life as if you have achieved your goals and desires. It's a mental rehearsal that helps align your subconscious mind with the reality you want to create. When you script, you write in the present tense, describing your ideal life in vivid detail, as though it's already happening.

Both journaling and scripting are powerful because they engage your mind in the creative process. By writing your thoughts and desires, you clarify what you want and reinforce your belief in the possibility of those desires becoming reality. These practices help you focus your energy and intention, making it easier to achieve your goals.

How to Practise Journaling and Scripting

Here's how you can integrate journaling and scripting into your daily routine to support your quantum manifestation journey:

1. **Set Aside Time Daily**: Dedicate a specific time for journaling or scripting each day. Morning or evening works best, as these times allow for reflection and intention-setting.

2. **Start with Journaling**: Write about your thoughts, feelings, and experiences. Reflect on your day, your challenges, and your successes. Use this time to unload any mental clutter and gain clarity.

3. **Transition to Scripting**: After journaling, shift to scripting. Write about your goals and desires as if you have already accomplished them. Use the present tense and describe your

ideal life. For example, if you want to find your dream job, you might write: 'I am thrilled to be working in my dream job, where I am valued, fulfilled, and excited to contribute every day.'

4. **Engage Your Emotions**: As you script, connect emotionally with the words you're writing. Feel the joy, satisfaction, and gratitude of living your dream life. This emotional connection amplifies the power of your scripting.

5. **Be Consistent**: Like all manifestation tools, consistency is vital. Make journaling and scripting a daily habit, and watch as your intentions manifest.

Taylor Swift, the globally acclaimed singer-songwriter, has often spoken about how journaling is integral to her creative process. From a young age, Taylor used journaling to work through her emotions, capture her thoughts, and explore her inspirations. This practice helped her process her experiences and served as a wellspring of ideas for her music.

Many of Taylor's songs are born out of the pages of her journals, where she writes her thoughts and emotions, turning them into lyrics that resonate with millions. By writing about her feelings and experiences, she can channel her emotions into her art, creating deeply personal yet universally relatable songs. Taylor's commitment to journaling has played a significant role in her ability to connect with her audience on such a profound level.

Kobe Bryant: The Power of Journaling

Kobe Bryant, the late NBA superstar and entrepreneur, was renowned for his relentless drive and focus, both on and off the court. One tool Kobe used to maintain this focus was journaling. He would document his thoughts, goals, and experiences as a professional athlete and businessman, using his journals to reflect on his performances and track his progress and strategies for the future.

Kobe's journaling practice was not just about documenting his achievements; it was a way for him to stay disciplined and mentally prepared. By writing his goals and visualising his success, Kobe reinforced his commitment to excellence, which translated into his legendary performances on the basketball court and his successful ventures in business after he retired from sports.

These examples should provide a fresh and powerful illustration of how Kobe Bryant and Taylor Swift used journaling and scripting to manifest success.

The Power of Writing

Research has shown that writing can profoundly affect the brain and body. Writing engages multiple brain areas, including those responsible for thinking, language, and motor control. When you write about your goals, you're processing them mentally and physically, which helps solidify your intentions. Writing also helps to organise your thoughts, making it easier to focus on and achieve your goals.

Exercise: Start Your Scripting Journey

1. **Choose a Quiet Place:** Find a comfortable and quiet spot where you can write without distractions.
2. **Reflect and Journal:** Begin journaling about your thoughts, emotions, and experiences. Reflect on what's going well and areas you'd like to improve.
3. **Shift to Scripting:** After journaling, start scripting your future as if your desires have already come true. Write in the present tense, describing your life in vivid detail.
4. **Feel the Emotions:** As you write, immerse yourself in the emotions of already achieving your goals. Let the excitement, gratitude, and joy flow through your words.
5. **Repeat Daily:** Make this practice a daily habit. Over time, you'll notice a shift in your mindset and a stronger alignment with your desired reality.

The Power of Meditation – Cultivating Stillness for Quantum Connection

"The stiller you are, the calmer your life is."

– Paramahansa Yogananda

Meditation is a practice that transcends the boundaries of time and culture. It's a tool for quieting the mind, cultivating inner peace, and connecting with the deeper aspects of our lives through the realm of quantum manifestation. Meditation plays a crucial role in aligning thoughts and emotions with the quantum field, creating a direct link between your desires and the universe's infinite potential.

What Is Meditation?

Meditation is the practice of focusing the mind and eliminating distractions to achieve a state of heightened awareness and inner calm. It involves training the mind to remain present, allowing you to connect with your true intentions and the underlying energy of the quantum field. Meditation can take many forms, from mindfulness and breath awareness to more advanced practices like transcendental meditation and guided visualisation.

In quantum manifestation, meditation serves as a gateway to the quantum field. By quieting the chatter of the conscious mind, you can access the deeper layers of your subconscious, where your genuine desires and intentions reside. This stillness also allows you to connect with the universe's energy, amplifying your ability to manifest your goals and dreams.

Types of Meditation

1. **Mindfulness Meditation**: This meditation focuses on the present moment, often by concentrating on the breath. It helps cultivate awareness and reduce stress by anchoring the mind in the here and now.

2. **Guided Visualisation**: In this practice, a guide leads you through visualisation, helping you to create the life you desire

mentally. This form of meditation is potent in quantum manifestation, combining meditation with vivid imagery and visualisation.

3. **Transcendental Meditation**: This technique involves silently repeating a mantra to help the mind settle into profound rest and heightened awareness. It's known for reducing stress and promoting mental clarity.

4. **Loving-Kindness Meditation**: Also known as *Metta* meditation, this practice focuses on developing compassion and love, first for oneself and then for others. It's a powerful way to cultivate positive energy and enhance emotional well-being.

Paramahansa Yogananda, the Indian yogi who introduced millions of Westerners to the teachings of meditation and Kriya Yoga, is a profound example of how meditation can connect one with the quantum field. Yogananda's teachings emphasised the importance of deep, focused meditation to tap into the divine and manifest one's highest potential.

Yogananda believed that, through meditation, individuals could access the universe's infinite potential and align themselves with their true purpose. His own life was a testament to this belief. Through his deep meditative practices, Yogananda manifested extraordinary outcomes, such as establishing the Self-Realisation Fellowship in the United States and inspiring millions through his writings, notably his autobiography, *Autobiography of a Yogi*.

Meditation and the Brain

Research has shown that meditation can significantly affect the brain, including increased grey matter in areas related to learning, memory, and emotional regulation. Studies show that meditation reduces the activity of the brain's default mode network (DMN), which handles mind-wandering and self-referential thoughts.

By meditating, this network is quieted, cultivating a state of focused awareness and making it easier to connect with the quantum field and manifest one's intentions.

Exercise: Cultivating Stillness Through Meditation

1. **Find a Quiet Space**: Choose a place where you can sit comfortably without distractions. It could be a corner of your home, a peaceful natural spot, or anywhere you feel relaxed.

2. **Focus on Your Breath**: Close your eyes and focus on your breath. Notice the sensation of air entering and leaving your body. If your mind wanders, gently bring your focus back to your breath.

3. **Set an Intention**: Before you begin, set a clear intention for your meditation. This could be a goal you're working towards, a state of being you wish to cultivate, or a desire for inner peace.

4. **Embrace the Stillness**: As you meditate, allow your mind to become still. Let go of any thoughts or distractions; be present in the moment.

5. **Practice Regularly**: Aim to meditate for 10-15 minutes daily. Over time, you'll find that your ability to connect with the stillness within you deepens, enhancing your capacity to manifest your desires.

The Power of Self-Love – The Foundation of Manifestation

"You, as much as anybody in the entire universe, deserve your love and affection."

– Buddha

Self-love is the cornerstone upon which we build other forms of manifestation. It is the practice of nurturing, accepting, and valuing yourself as you are. When you cultivate self-love, you align yourself with the universe's energy, allowing you to manifest your desires with greater ease and authenticity. Without self-love, the desires we seek may remain out of reach as the universe responds to our thoughts and the energy we hold within ourselves.

What is Self-Love?

Self-love is the conscious practice of caring for your physical, emotional, and mental well-being. It involves setting healthy boundaries, embracing your worth, and treating yourself with kindness and compassion. Self-love is not selfish; it's recognising your inherent value and ensuring that your actions and thoughts reflect that recognition.

In quantum manifestation, self-love is essential because it sets the tone for what you believe you deserve. When you love yourself, you send a powerful signal to the universe that you are worthy of your desires. This belief in your worthiness attracts positive experiences, relationships, and opportunities that align with your highest good.

How to Cultivate Self-Love

Cultivating self-love is a lifelong journey that requires consistent practice and commitment. Here are some steps to help you develop and nurture self-love:

1. **Practise Self-Compassion**:
   Treat yourself with the same kindness and understanding you would offer a close friend. When you make a mistake or face a challenge, avoid harsh self-criticism. Instead, acknowledge your feelings and offer yourself words of encouragement.

2. **Set Healthy Boundaries**:
   Learn to say no to situations, people, or activities that drain your energy or compromise your well-being. Setting boundaries is an act of self-respect and helps to protect your emotional and mental health.

3. **Engage in Self-Care**:
   Make time for activities that nourish your body, mind, and soul, including exercise, meditation, time in nature, or simply doing things that bring you joy.

4. **Embrace Your Worth**:
   Recognise and affirm your value daily. Use positive affirmations to reinforce your belief in your worthiness. Remind yourself that you deserve love, respect, and happiness.

5. **Forgive Yourself**:
   Let go of past mistakes and regrets. Understand that you are human, and mistakes are part of the learning process. Forgiving yourself frees you from guilt and allows you to move forward with greater self-compassion.

Real-Life Examples of Self-Love

Kamal Ravikant, the author of *Love Yourself Like Your Life Depends On It*, experienced a life-altering crisis that pushed him to the edge. In despair, Kamal realised that radical self-love was the only way to overcome his challenges. He began affirming love for himself daily, even when it felt difficult or unnatural.

Over time, this practice of self-love transformed Kamal's life. It brought him out of depression and into a state of inner peace and resilience. Kamal's journey shows the power of self-love to heal and empower, making it a critical tool for anyone looking to manifest positive change.

Louise Hay, the renowned author of *You Can Heal Your Life*, used self-love as a powerful tool to overcome significant challenges, including her battle with cancer. Diagnosed with an incurable illness, Louise turned to the practice of daily affirmations and self-love as part of her healing journey.

Louise's holistic approach combined positive affirmations, visualisation, and self-care with traditional medical treatments. By consistently practising self-love and reinforcing her belief in her worth, Louise overcame her illness and, through her teachings and books, helped millions of others do the same.

Louise Hay's story is a testament to the transformative power of self-love. It shows how loving yourself can lead to profound healing and open the door to excellent health, happiness, and abundance.

The Psychology of Self-Love

Psychological research supports that self-love is crucial for mental and emotional well-being. Studies have shown that individuals who practise self-compassion and self-love have higher self-esteem, better coping mechanisms, and lower stress and anxiety levels. This positive self-regard creates a fertile ground for personal growth and manifestation, as it encourages a mindset of abundance rather than scarcity.

Exercise: The Mirror of Self-Reflection

1. **Find a Mirror:**
   Stand in front of a mirror where you can see your entire reflection.

2. **Speak Words of Love:**
   Look into your eyes and say, "I love you," and, "I am worthy of all good things." Repeat these affirmations several times, allowing the words to resonate within you.

3. **Acknowledge Your Worth:**
   As you speak these words, focus on the emotions that stir within you. If feelings of doubt or resistance arise, acknowledge them without judgement and gently reaffirm your commitment to self-love.

4. **Practise Daily:**
   Make this mirror exercise a daily habit. Over time, you'll find that your belief in your worth strengthens, and the energy you project to the universe becomes more aligned with your desires.

5. **Extend Compassion to Yourself:**
   Remember to be gentle with yourself throughout the day, especially during challenging moments. This ongoing practice of self-love will enhance your ability to manifest your goals and dreams.

Quantum Leaping and Parallel Realities – Shifting into Your Desired Life

"The reality you live in is just one of the many outcomes you could experience."

— Vadim Zeland

Quantum leaping and parallel realities are concepts rooted in the idea that infinite possibilities exist in our endless universe, and our current reality is just one of many that exist simultaneously. In the realm of quantum manifestation, these ideas suggest that by shifting our focus, energy, and beliefs, we can leap from one reality to another—specifically into the reality that best aligns with our desires and goals.

As the intricacies of quantum leaping unfold in a dedicated chapter later in this book, I shall limit myself to offering the practical method—how to step into the art of quantum leaping. In that later chapter, the concept is explored in depth, defined with clarity, and illuminated with insight, for it felt fitting that such a vital aspect of quantum creation deserves a space of its own.

The Observer Effect and Quantum Leaping

This concept closely relates to the observer effect in quantum physics, which suggests that observation can influence a quantum event's outcome. Similarly, your focus and intention can determine which reality you experience. By quantum leaping, you are essentially 'observing' the reality you want and pulling it into your experience.

What are Parallel Realities?

Parallel realities refer to the idea that multiple versions of reality exist side by side, each representing different outcomes and possibilities. Every decision creates a new branch of reality, leading to a different result. In this vast multiverse, there are infinite versions of you, each living out different scenarios based on the choices you have made or could make.

The idea of parallel realities can be exciting and empowering, as it suggests that your current circumstances do not bind you. Instead, you can shift into a reality that aligns better with your genuine desires by changing your focus, beliefs, and actions.

Vadim Zeland and Reality Transurfing

Vadim Zeland, the Russian author and mystic, introduced the concept of 'Reality Transurfing' in his groundbreaking book series. Zeland's work centres on the idea that we live in a space of variations—a quantum field where all realities exist simultaneously. According to Zeland, we can 'surf' through these variations by consciously directing our thoughts and emotions and choosing the reality we wish to experience.

In *Reality Transurfing*, Zeland emphasises the importance of intention and awareness in navigating the quantum field. He suggests that by maintaining a clear vision of your desired reality and staying detached from the outcome, you can move effortlessly through distinct realities until you land the one that aligns with your desires. Zeland's teachings provide a practical guide for anyone looking to consciously create their reality and make quantum leaps into their ideal life.

How to Practise Quantum Leaping

Quantum leaping is a process that requires a combination of focus, belief, and action. Here's how you can practise quantum leaps in your own life:

1. **Define Your Desired Reality**:
   Clearly define the reality you want to experience. Write it down, describing every aspect of this new reality as vividly as possible.

2. **Visualise and Feel**:
   Daily, visualise yourself living in this new reality. Engage all your senses and feel the emotions associated with this experience. The more natural it feels, the stronger your connection to this reality will become.

3. **Align Your Actions:**
    Align your actions with your desired reality. This might involve changing your daily habits, mindset, or environment to reflect your desires.

4. **Detach from the Outcome:**
    One of the fundamental principles of quantum leaping is detachment. While focusing on your desired reality is essential, remaining unattached to the specific outcome is equally important. Trust that the universe will bring you to the right reality in its own time.

5. **Practise Gratitude:**
    Cultivate gratitude for the reality you are creating, even before it fully manifests. Gratitude helps reinforce your connection to this new reality and speeds up the process of quantum leaping.

Scientific Support for Parallel Realities

The concept of parallel realities gets support from the multiverse theory in physics, which posits that our universe is just one of many universes that exist simultaneously. These universes could have different physical laws, outcomes, and realities. While the multiverse theory is still debated among scientists, it provides a fascinating framework for understanding how quantum leaping might work in quantum manifestation.

Exercise: Making a Quantum Leap

1. **Set the Stage:**
    Find a quiet place to relax and focus. Close your eyes and take several deep breaths to calm your mind.

2. **Visualise Your Leap:**
    Picture yourself in your current reality. Imagine a portal or doorway in front of you that leads to your desired reality. Visualise yourself stepping through this portal into your new life.

3. **Engage with Your New Reality**:
   Take in every detail as you enter this new reality. What does it look like? How does it feel? Spend several minutes immersing yourself in this new reality, allowing yourself to experience it fully.

4. **Anchor the Leap**:
   When you're ready, open your eyes and bring the feeling of your new reality with you. Throughout the day, remind yourself that you live in this new reality. Let your thoughts, actions, and decisions reflect this shift.

5. **Repeat Regularly**:
   Make quantum leaping a regular part of your manifestation practice. The more you practise, the easier it will become to shift into your desired reality.

Healing the Inner Child, Embracing the Shadow, and Reprogramming the Subconscious Mind

"The child within us is the seed of our future potential, and the shadow is the soil in which it grows."

– Carl Jung

To manifest your desired reality, you must address the deeper layers of your psyche—those parts of yourself that may hold onto past wounds, limiting beliefs, and unresolved emotions. Healing the inner child, embracing the shadow, and reprogramming the subconscious mind are potent practices that help clear these blockages, making it easier for your desires to manifest in the physical world.

Healing the Inner Child

The inner child is the part of your psyche that retains the innocence, wonder, and emotions of your early years. This inner child also carries the wounds and traumas of childhood, which can manifest as limiting beliefs, fears, and emotional triggers in adulthood. Healing the inner child is about reconnecting with this vulnerable part of yourself, offering love, understanding, and reassurance.

When you heal your inner child, you release the emotional blockages holding you back. This process allows you to reclaim your childhood joy, creativity, and spontaneity, integrating these qualities into your adult life. By nurturing your inner child, you create a solid foundation for quantum manifestation as you free yourself from the past and open up to new possibilities.

Embracing the Shadow

The shadow, a concept introduced by Carl Jung, represents the unconscious aspects of your personality – those parts of yourself that you may reject, deny, or suppress. These aspects often include traits and emotions that society deems undesirable, such as anger, jealousy, or fear. However, the shadow also contains vital energy and potential that, when integrated, can lead to profound transformation.

Shadow work brings these unconscious aspects into the light of awareness, embracing them with compassion and understanding. By acknowledging and integrating your shadow, you reclaim the energy locked away, using it to fuel your growth and manifestation journey. Embracing the shadow is essential for achieving wholeness and aligning your entire being with your desired reality.

Reprogramming the Subconscious Mind

The subconscious mind is the powerhouse of your beliefs, habits, and automatic responses. It operates below the level of conscious awareness, influencing your thoughts, emotions, and behaviours. Many beliefs stored in the subconscious mind are formed in childhood and may no longer serve you as an adult. Reprogramming the subconscious mind involves replacing these outdated beliefs with new, empowering ones that support your goals and desires.

Several techniques exist for reprogramming the subconscious mind, including self-hypnosis, autosuggestion, and affirmations. The key to effective reprogramming is repetition and emotional engagement. By consistently feeding your subconscious mind with positive, empowering beliefs, you gradually overwrite the old programming, aligning your subconscious with your conscious intentions.

Carl Jung, one of the most influential figures in psychology, deeply explored the concept of the shadow in his work. Jung believed that integrating the shadow was crucial for achieving wholeness and self-actualisation. He argued that by confronting and embracing the darker aspects of our psyche, we could unlock hidden potential and live more authentic lives.

Jung's journey involved extensive shadow work, where he confronted his unconscious fears, desires, and impulses. This process allowed him to gain profound insights into the human psyche, enhancing his self-awareness and contributing to his groundbreaking theories in psychology. By embracing his shadow, Jung could harness the energy that had been repressed, and he used it to fuel his creative and intellectual pursuits. His work on the shadow continues to influence psychological practices today, emphasising the importance of integrating all aspects of the self for actual personal growth and transformation.

Jung's exploration of the shadow teaches us that the parts of ourselves we most fear or reject can become powerful allies in our journey towards wholeness. By working to understand and integrate these aspects, we open the door to a more prosperous, more authentic life and align ourselves more fully with the realities we wish to create.

Napoleon Hill, author of the classic self-help book *Think and Grow Rich*, emphasised reprogramming the subconscious mind as the key to success. Hill introduced the concept of autosuggestion, a technique where individuals repeatedly affirm their goals and desires to themselves, effectively programming their subconscious mind to believe in their inevitable success.

Hill's own life reflected the principles he taught. Through persistent use of autosuggestion, Hill overcame significant obstacles, including poverty and self-doubt, to become one of the most influential motivational writers of the 20th century. His teachings on the power of the subconscious mind have inspired countless people to achieve their dreams by reprogramming their thoughts and beliefs to align with their goals.

Hill's story shows the power of the subconscious mind in shaping our reality. By consciously feeding the mind with positive, empowering beliefs, you can shift your internal programming, making it easier to manifest your desires in the external world.

How to Practise Inner Child Healing, Shadow Work, and Subconscious Reprogramming

Here are some practical steps to help you engage in these transformative practices:

Inner Child Healing

1. **Connect with Your Inner Child**: Find a quiet place to close your eyes and visualise yourself as a child. Picture the younger version of yourself, and observe any emotions or memories that arise.

2. **Offer Love and Reassurance**: Speak to your inner child with kindness and compassion. Reassure them that they are safe, loved, and valued. Acknowledge any pain or fear they might carry, and offer comfort.

3. **Incorporate Play:** Engage in activities your inner child enjoys, such as drawing, playing games, or spending time in nature. This will help rekindle the joy and creativity of your childhood.

4. **Journal Your Experience:** Keep a journal about your experience after each session. Write down any insights or emotions that surfaced during your inner child work and reflect on how you can continue to nurture this part of yourself.

Shadow Work

1. **Identify Your Shadows:** Reflect on the traits or emotions you find difficult to accept in yourself. These might include feelings of anger, jealousy, or fear. Acknowledge these aspects without judgement.

2. **Create a Dialogue:** Write a conversation between yourself and your shadow. Ask your shadow what it needs and listen to its response. This process helps you understand and integrate these parts of yourself.

3. **Embrace Your Shadow:** Instead of rejecting or suppressing your shadow, embrace it as part of who you are. Recognise the strengths and potential that lie within these aspects of yourself.

4. **Seek Support:** Shadow work can be challenging, so consider seeking guidance from a therapist or counsellor, especially if you encounter deeply buried emotions or traumas.

Reprogramming the Subconscious Mind

1. **Use Affirmations:** Create positive affirmations that reflect the beliefs you want to instil in your subconscious mind. Repeat these affirmations daily, especially in the morning and before bed, when the subconscious is most receptive.

2. **Practise Self-Hypnosis:** Find a quiet space, relax deeply, and guide yourself into a state of focused attention. In this state, repeat your affirmations or visualise your desired outcomes, allowing them to sink into your subconscious mind.

3. **Visualise Success:** Spend time each day visualising yourself as having already achieved your goals. Engage all your senses in the visualisation and feel the emotions associated with your success.

4. **Be Consistent:** Reprogramming the subconscious mind takes time and repetition. Stay consistent with your practices, and you'll notice a shift in your beliefs and behaviours.

The Power of Collective Consciousness in the Quantum Circle

"Alone, we can do so little; together, we can do so much."

– Helen Keller

In the vast expanse of the quantum field, where thoughts are energy and possibilities are infinite, there lies an incredible power—one that transcends individual efforts and taps into the collective mind of humanity: the power of collective consciousness, a force that, when harnessed, can shape the reality of not just one person but of communities, societies, and even the world.

The Concept of Collective Consciousness

Collective consciousness refers to the shared beliefs, ideas, and moral attitudes that operate as a unifying force within society. It is the invisible thread that connects us all, weaving individual thoughts and emotions into a larger tapestry of shared reality. In quantum manifestation, collective consciousness acts as a powerful amplifier, turning the isolated whispers of individual intentions into a resonating chorus that can bring about profound changes in the material world.

The Quantum Circle: Harnessing the Power Together

Imagine the Quantum Circle as a gathering of minds, all focused on a singular intention. Within this circle, individuals align their thoughts, emotions, and energies towards a common goal, creating a synergy far exceeding the sum of its parts. The Quantum Circle is a space where the power of collective consciousness is consciously cultivated and directed, much like a laser beam that cuts through the chaos of the quantum field to manifest a specific outcome.

Each participant brings unique energy and perspective in a Quantum Circle, yet a shared vision unites them all. Whether it's a group meditation, a synchronised global event, or even a shared intention within a family or team, the effects of such collective focus can be extraordinary. Scientific studies have shown that group meditation can reduce crime rates, lower stress, and even influence global events—a testament to the power of collective consciousness.

One remarkable example of the power of collective consciousness is the *Maharishi Effect*, a phenomenon named after Maharishi Mahesh Yogi. In the 1970s and 1980s, studies conducted on large groups practising transcendental meditation showed that the overall quality of life improved significantly when a critical mass of people meditated on peace and harmony. Crime rates dropped, social coherence increased, and economic indicators showed positive shifts. These changes were not mere coincidences; they were collective consciousness in action, altering the fabric of reality.

Another modern example is the global *Intention Experiments* led by author Lynne McTaggart. These experiments have involved thousands of people worldwide, focusing their thoughts on specific outcomes—such as healing a person, improving crop yields, or reducing violence in a conflict zone. The results, often measured scientifically, have shown that collective intention can influence reality, demonstrating the potential of collective consciousness in quantum manifestation.

How to Create Your Own Quantum Circle

Creating your own Quantum Circle is simple yet profound. Start by gathering a group of like-minded individuals who share a common goal. This goal could be as small as a weekly gathering with friends or as large as a global online community. Establish a clear intention for the circle, ensuring everyone aligns and commits to the shared vision.

Focus your thoughts and energies on the desired outcome during your gatherings, whether in person or virtually. This could involve group meditations, visualisation exercises, or even synchronised actions in the physical world. The key is maintaining a high vibration, focusing on positive emotions, and holding an unwavering belief in the group's collective power.

The Ripple Effect of Collective Consciousness

When individuals come together in a Quantum Circle, the impact of their collective consciousness can create a ripple effect that extends far beyond the immediate goal. It can inspire others, foster a sense of unity, and even contribute to the evolution of human consciousness globally.

In the realm of quantum manifestation, the power of one is great, but the power of many is unstoppable. By harnessing the collective consciousness within a Quantum Circle, we tap into a wellspring of potential that can manifest the most extraordinary realities—for ourselves and the world.

Reflective Questions

- How can you harness the power of collective consciousness in your own life?
- What shared goals can you pursue with others in a Quantum Circle?
- How might participating in or leading a Quantum Circle change your approach to manifestation?

Connect Online

1. **Join Forums or Social Media Groups:** Connect with people sharing experiences in quantum manifestation.
2. **Engage Locally:** Attend workshops, seminars, or group meditations to meet like-minded individuals.
3. **Share Your Journey:** Whether through blogging, vlogging, or conversations, sharing your experiences can inspire others and strengthen your practice.

Many readers might wonder whether they should focus on a single approach or combine methods for better results. I wish to address these questions to help clarify the path for those seeking to master quantum manifestation.

One advantage of integrating multiple techniques is that it allows for a more comprehensive approach to manifestation. For instance, combining visualisation with affirmations ensures that your conscious and subconscious minds align with your goals. Visualisation helps you see and feel your desired outcome as if it's already happening, while affirmations reinforce these positive images with supportive thoughts, reprogramming your subconscious mind to accept these new realities.

Many books continue the discussion of quantum tools in this chapter and delve deeper into these and related techniques, offering readers a broader understanding of quantum manifestation. For those looking to expand their exploration of these concepts, the following books provide invaluable insights and practical guidance to enhance their journey.

Recommended Reading

One such book is *The Holographic Universe* by Michael Talbot, which explores the idea that the universe is a hologram where every part contains the whole, much like how our consciousness interacts with the quantum field. Talbot's work is relevant for understanding the interconnectedness of all things and how our beliefs and perceptions shape the reality we experience.

This concept complements the practice of shadow work and inner child healing, where integrating unconscious aspects of ourselves can lead to profound shifts in our external reality. Talbot's exploration of quantum energy and the nature of reality provides a framework for understanding how our inner world reflects in the outer world—a core principle of quantum manifestation.

Another essential read is *Breaking the Habit of Being Yourself* by Dr. Joe Dispenza, which combines neuroscience with quantum physics to explain how we can rewire our brains and change our lives. Dispenza focuses on the power of meditation, visualisation, and intention-setting to create new neural pathways that align with the reality we wish to manifest.

His work highlights that consistently practising these techniques allows us to break free from old patterns and create a new, more fulfilling reality. This aligns with quantum leaping and reprogramming of the subconscious mind, where deliberate practice and focus can significantly change one's life.

Quantum Healing by Deepak Chopra integrates quantum principles with health and well-being, showing how self-love and meditation can influence physical reality by aligning mind and body through the quantum field.

Ask and It Is Given by Esther and Jerry Hicks, based on the teachings of Abraham Hicks, emphasises the importance of aligning your thoughts and emotions with the frequency of what you desire. The book introduces various practical tools, such as the Emotional Guidance System, which helps individuals navigate their feelings to maintain a high vibrational state conducive to manifesting their desires.

This concept ties into the law of attraction and the power of intention, both essential elements in quantum manifestation. By understanding and applying these tools, readers can learn to control their energy better and direct it towards achieving their goals.

'The Power of Intention' by Dr. Wayne Dyer further expands the concept of intention as a creative force. Dyer emphasises that when we set clear, focused intentions and align our energy with those intentions, we become co-creators with the universe. His approach underscores the importance of maintaining a high vibrational state and using visualisation and affirmations to keep your intentions at the forefront of your consciousness. Dyer's teachings benefit those looking to harness the power of intention as a central component of their manifestation practice.

For readers interested in the collective aspect of quantum manifestation, 'The Field: The Quest for the Secret Force of the Universe' by Lynne McTaggart offers a compelling exploration of how individual thoughts are part of a more extensive, interconnected web of energy. McTaggart provides evidence that collective thought and intention can influence reality on a broader scale, supporting the concept of collective consciousness.

This book is a valuable resource for those interested in participating in or creating Quantum Circles, where a group's combined energy and focus can amplify the effects of individual manifestations. *Eckhart Tolle's 'The Power of Now'* emphasises the importance of present-moment awareness. This practice allows you to tap into the quantum field's potential and leap into new realities by focusing on the now.

The work of pioneers like **Napoleon Hill** and **Louise Hay** adds a layer of practicality to these metaphysical concepts. Hill's *Think and Grow Rich* remains a cornerstone of success literature, emphasising the power of desire, faith, and persistence—elements mirrored in quantum theory's focus on the observer effect and the role of intention in shaping reality.

Louise Hay offers a deeply personal testament to the power of affirmations and positive thinking. Diagnosed with cancer, Hay turned to the practice of daily affirmations and self-love, a journey she chronicles in her book *You Can Heal Your Life*. Her story is a powerful example of how changing one's thoughts and beliefs can lead to profound healing and transformation, both physically and spiritually.

Lastly, *'The Dancing Wu Li Masters'* by Gary Zukav offers a deep dive into the intersection of quantum physics and consciousness, providing a philosophical and scientific foundation for understanding how our thoughts and perceptions shape reality. Zukav's work is essential for those who wish to understand the underlying principles of quantum manifestation, particularly the role of the observer in determining outcomes. This book is a perfect companion for readers exploring the scientific basis of the metaphysical concepts discussed in this chapter.

Further, movies like *'What the Bleep Do We Know!?' and 'The Secret'* take these concepts from the page to the screen, blending science, spirituality, and narrative to explore the impact of consciousness on reality. The films present quantum manifestation as a philosophical idea and a lived experience, showcasing how our thoughts and beliefs shape our world.

Fictional characters such as *'Harry Potter' and 'Luke Skywalker'* embody these principles in their journeys. Harry's ability to summon his Patronus, a powerful protective force, and Luke's mastery of the Force vividly illustrate how belief and focus can transform reality. These characters, though fictional, resonate with real-world applications of quantum principles, serving as allegories for the power we each hold to shape our destiny.

Neuroscience further supports these ideas by demonstrating how the brain's plasticity allows rewiring through repeated thought patterns and affirmations. This scientific perspective provides a tangible framework for understanding how visualisation and affirmations work at a neurological level, reinforcing the belief that we can shape our realities.

The intersection of quantum mechanics, neuroscience, and metaphysical practices is not just an abstract concept—it is a powerful testament to the potential within each of us to manifest the life we desire. As we immerse ourselves in these books and films, we gain not only knowledge but also the inspiration to apply these principles in our lives, transforming theory into practice and imagination into reality.

These books and movies collectively offer a wealth of knowledge and practical techniques that can complement and enhance the quantum tools discussed in this chapter. By exploring these works, readers can deepen their understanding of how to use visualisation, affirmations, intention, and other quantum techniques to consciously create the reality they desire. Whether you are just beginning your journey or looking to refine your practice, these books provide valuable insights and guidance to help you navigate the complex yet profoundly rewarding path of quantum manifestation.

Conclusion: Crafting Your Quantum Masterpiece

The tools of visualisation, affirmations, meditation, self-love, inner child healing, and shadow work are your brushes and paints on the grand canvas of life. As you integrate these practices into your daily routine, you'll notice a subtle yet profound shift in your perception of reality. The world around you will reflect the intentions you set, the beliefs you hold, and the love you cultivate within yourself.

But remember, this is not a journey with a final destination; it is an ongoing process of creation and transformation. With each step, you become more attuned to the infinite possibilities of the quantum field and more aware of your power as a co-creator of your reality. Trust, embrace the unknown, and allow the universe to surprise you with its boundless potential.

As you continue to explore the depths of your consciousness, may you find the courage to dream greater dreams, the wisdom to trust the journey, and the love to nurture yourself along the way. The quantum field is your playground; the only limit is the one you set for yourself. So craft your quantum masterpiece, one thought, one belief, one action at a time.

CHAPTER 13

Overcoming Common Obstacles and Eliminating Myths

Through the open windows of my writing hideaway, the cool Nilgiri breeze gently flowed. The soft sunlight stroked the worn pages of my journal, illuminating the ink in a reassuring way. The distant mountains stood tall and majestic, their summits crowned with a colourful canopy of trees, bathed in the soft light of dawn. Immersed in this tranquil setting, with the soft rustle of pine trees and the ethereal chirp of birdsong filling the air, I couldn't help but think about the difficulties we face when trying to realise our ambitions.

When subtle chances get overshadowed by overpowering pessimism, even the most dedicated person may face moments of uncertainty. In these moments, we can overcome mental hurdles and refocus our attention on our goals by using quantum physics and the Law of Assumption.

Unexpectedly, doubt can arise and pose a significant obstacle. It insinuates itself into our minds, weaving tales of failure and past letdowns. There will be times when you wonder if you are worthy of your dreams and if you have what it takes to achieve them. As a result of past disappointments, this uncertainty follows you wherever you go.

Exciting and thought-provoking in quantum physics is the observer effect. Think about the person observing in the double-slit experiment; their very presence changed how the light particles behaved.

Even when our thoughts fill us with doubt, we can construct our reality as mere witnesses to the grand cosmic drama.

Imagine your mind as a vibrant garden. Dreams are like fragile flowers; uncertainty has a way of sneaking in and suffocating them. Like a skilled gardener tending to their plants, we may eliminate uncertainty by focusing on our intent. Allow me to propose an easy method that has worked well for me:

A Way to Remove Any Doubt

1. **Acknowledge Its Presence:** Pay attention to that nagging uncertainty. 'Hey there, doubt. Seems like you've paid me a visit'. This is a polite way to acknowledge uncertainty.
2. **Shift Your Perspective:** Rather than letting uncertainty consume you, take a moment to reflect on the deep-seated anxieties that could drive it. Perhaps it's simply a fear of falling short or a lack of self-worth. Who can say for sure? You can overcome your fear if you know where it's coming from.
3. **Use Affirmations:** Affirmations that support your goals can help you overcome the negative thoughts that accompany doubt. Swap out pessimistic ideas for optimistic and empowering ones. For example, if you want to gain a new job, stop second-guessing yourself and start believing that you deserve your ideal role instead of doubting your qualifications. Hold steadfastly to these affirmations, letting their indisputable truth sink in.

It is a common experience for many people to suffer from feelings of loneliness and emptiness. It might be challenging to accomplish our objectives when we encounter challenges such as a lack of resources, relationships, or unsteady resolve. Quantum physics offers an exquisite resolution to a widespread misperception about reality, leaving one with a profound sense of possibility.

The cosmos is a gigantic, predetermined machine in the classical Newtonian worldview. As with stones dropped into a still pond, desires cause ripples that eventually fade away. Quantum physics, however, offers another perspective. It conjures up the image of a world without boundaries—a boundless ocean of outcomes. Our dreams and ambitions illuminate the path to the results we seek, driven by an inner sense of purpose.

Here's a riveting tale. I will tell you the story of **Jack Andraka**, a unique man. What's most interesting about this extraordinary individual is that he accomplished his significant discovery without the aid of a posh lab or a team of scientists. With nothing more than a computer and an insatiable thirst for knowledge, he achieved his breakthrough.

What, then, was his discovery? It was a major advancement in the early detection of cancer. Isn't that remarkable? A young man who didn't seem to have much money or powerful connections made a groundbreaking scientific discovery thanks to his dogged persistence and unshakeable self-assurance.

Feeling inadequate is nothing to worry about. Our connection to the cosmos is deep and meaningful. With focused intention and active participation, we have the power to shape reality. To help you overcome feelings of lack, I have a simple exercise:

Abundance Exercise

1. **Find a Quiet Spot:** Retreat to a peaceful place for a moment. Unwind and find tranquillity within.

2. **Visualise Your Ideal Reality:** Envision a universe where all your wishes have come true. Picture yourself strolling through the spacious, light-filled rooms of your brand-new home. Imagine commanding the respect of your colleagues as you expertly lead a meeting. It's as if your dreams have already materialised.

3. **Embrace the Emotions:** Immerse yourself in the exhilarating emotions linked to your wildest dreams. Give in to the overwhelming joy, ease, and profound contentment. Remain in this state of mind for as long as you can without feeling rushed.

With regular practice, we can overcome even the most insurmountable mental obstacles to manifestation. Remember that the cosmos is entirely unconcerned with our uncertainties and fears.

The concept of time can be quite perplexing when we try to grasp it. The details of when our wishes will be fulfilled can become a source of obsession as we eagerly await their realisation. This fixation not only causes frustration but also prevents us from seizing the wonderful opportunities available to us.

However, quantum physics offers a more flexible perspective on time. It alludes to the idea of a **'block universe'**, where past, present, and future coexist simultaneously. This challenges our linear perception of time, revealing instead a complex web of interrelated possibilities.

Imagine this: you plant a seed in the earth and resist the urge to dig it up every so often to check on its progress. You have faith in the laws of nature and know that the seed will grow into a beautiful flower with enough time and attention. Similarly, manifestation works the same way. We sow the seed of our desire with intention, nurture it with positive emotions and thoughts, and trust that the universe will bring it to fruition at the best possible time.

Visualisation Exercise

1. **Find a Quiet Space:** Go somewhere peaceful where you can unwind. Sit comfortably, close your eyes, and take a deep breath. Centre yourself.

2. **Visualise Your Desired Reality:** Picture your ideal vision clearly in your mind's eye as though it has already materialised. Enjoy the immense sense of achievement and joy that comes with reaching your goal.

3. **Bridge the Gap:** Imagine connecting the space between where you are now and where you want to be. Visualise your present circumstances merging with the ideal reality you aspire to live in. See yourself advancing step-by-step from your current state to your dream life by simply connecting the dots.

When we accept the natural flow of time, we can let go of the need for control. We trust that everything is happening at the perfect moment within the universe and believe that our dreams are woven into the vast fabric of reality.

Many individuals have trouble grasping the concept of manifestation and wrongly assume it has something to do with acquiring worldly goods. The pressure of constantly proving one's worth through acquiring more and more material goods or seeking approval from others can easily consume one. Genuine authenticity transcends merely gaining possessions. Completely embracing the person we must become to attract and embody our desires is of the utmost importance.

We are not helpless bystanders but active creators of our own world, as quantum mechanics and the Law of Assumption so starkly show. Each of us is unique, and because of this, the events and circumstances that speak to our authentic selves gravitate towards us.

If you want to change your mindset from acquiring to becoming, I have an embodying practice for you. To advance and evolve, it is fantastic to reflect and grow by scripting.

1. **Find a quiet place:** Take a few moments to relax and focus on your breathing.

2. **Feel Free to Express Your Desires:** Instead of fixating on getting the thing, try to embody the qualities that will bring what you want to you. Focusing on the qualities you want to develop in yourself should be your priority when starting a new romantic relationship. Think of admirable traits like self-assurance, empathy, and generosity.

3. Make sure the affirmations you make speak to who you really are and highlight your special attributes. As you confidently repeat these affirmations, welcome the profound change taking place inside you.

As we progress on our path of self-discovery, we inevitably attract people and situations that reflect our authentic selves. The focus is on boosting our personal development. Keep in mind that the point of manifesting is not to exert dominion over the cosmos. The important thing is to welcome the limitless possibilities that exist both within and outside of us.

The path to manifestation is rarely smooth, so keep that in mind as you embark on your journey. You will often encounter obstacles, moments of self-doubt, and the temptation to revert to old habits. You can overcome these obstacles and bring your chosen life into manifestation with constant effort, a caring perspective, and a readiness to embrace life's highs and lows. Everything in the cosmos is a living, evolving masterpiece that is just waiting for you to put your concentrated will into it. Embrace your inner artist, paint with conviction, and see your vision come to life.

As the sun dipped below the horizon in the Nilgiris, it created long shadows on the weathered pages of my journal. I heard the gentle rustling of pines accompanying the faint sound of crickets chirping in the distance and the aroma of damp soil. I closed my eyes and inhaled deeply, relishing the peace of the setting sun.

As we explore the ins and outs of making our dreams come true, a huge realisation hits us: if we want to reach our full potential, we need to become experts in our field and change our outlook completely. Manifestation is a skill that everyone may learn to use. It originates in our deep connection to the infinite possibilities of the universe and is, hence, an essential aspect of who we are.

Picture yourself gazing out across the breathtaking expanse of a magnificent mountain range. Looking out into the distance, you can make out a valley that seems to go on forever. Submerged in an enthralling mist, its splendour lies hidden. This valley embodies the boundless potential of the quantum world. There is an irresistible charm in the valley, where dreams and aspirations entwine, waiting for discovery.

We used to have it so easy to manifest, making all our dreams come true with little work at all. However, a more liberating insight may be revealed by adopting a quantum viewpoint. This valley holds a special place in our hearts; it is a place of boundless potential, and our lives will always be wrapped up in it.

The paths of the quantum field are inseparable from our mental processes, including our emotions, goals, and thoughts. By embracing a mindset that is in harmony with our aspirations, we play an active role in moulding our lives.

This transformation of awareness is an ongoing process, not a one-time occurrence. Okay, then, how about we look at a few ways to cultivate and enhance this heightened awareness:

Practice Gratitude: When you are grateful, you open yourself up to receiving more of what you desire. Pause for a moment and be grateful for the little things that come your way every day.

Enjoy the splendour of present-moment existence and cultivate the habit of paying close attention to your inner dialogue. Recognise when negative thoughts pop into your head and then gently refocus your attention on what you truly want. Staying focused and managing your thoughts can be a game-changer for reaching your goals.

Embrace the here and now; let go of the past and focus on the future. The only time for manifestation is the here and now. A limitless universe of possibilities opens up when one embraces the present-moment.

Keep in mind that you can't just insert a penny and get whatever you want out of the cosmos; it's not that simple. Emotions and ideas get woven into a complex web. Our growing knowledge weaves itself into the intricate fabric of life, impacting people and causing a chain reaction that goes far beyond our own hopes and dreams.

Masaru Emoto and his water crystals provide an interesting window into this narrative of a complex web of relationships. He made a surprising discovery about the profound impact of human intention and emotion on matter itself. Just think of the difference we could make if we all worked together to build a peaceful planet.

Manifestation is an exciting path of inner discovery. Exploring our full potential means welcoming the limitless opportunities that exist within us. With unfaltering determination, we should embrace the cosmic energies with all our beings and mould our desires accordingly.

To accompany you on this journey, keep in mind the wise words of physicist Werner Heisenberg: "Merely observing alters reality, as it interrupts the course of events." So, instead of being afraid of the quantum realm, embrace it with an attitude of boundless potential. Immerse yourself in the mesmerising observer effect, harness the tremendous potential of your concentrated intention, and behold the awe-inspiring realisation of your aspirations within the grand orchestra of life.

A Two-Faced Sword: Belief's Power

According to quantum physics and Goddard's teachings, belief matters in the manifestation process. A person's beliefs can both inspire and limit their actions; they can do both the good and the bad. The intriguing realm of the placebo effect in medicine is waiting for us. Amazingly, when patients have faith in the right treatment, they often see actual changes in their diseases. It is fascinating to see how our beliefs can actually influence how we perceive the world.

There is an interesting story regarding the placebo effect's incredible potency. Some patients who were experiencing knee discomfort took part in mock surgeries. They believed they had the actual surgery. Similar to those who have actual procedures, these patients had incredible pain relief. Obviously, their belief that a certain cure would result brought about the desired effect.

Getting Past Faith's Roadblocks

Understanding the limitations we set for ourselves is critical for releasing the full potential of our beliefs. Because of one's upbringing, social milieu, and level of self-confidence, one's origins can have far-reaching consequences. Think about how hard it is for many people to get ahead financially. Many people struggle with this issue because their upbringing taught them that wealth is scarce or that they are unworthy of it.

Task: Recognising Limiting Beliefs

1. Identify all the parts of your life that trigger problems and give them some thought. Jot down any mixed feelings or contradictory ideas you may have regarding these topics.
2. What are the fundamental assumptions behind these thoughts? Find out by digging deeper. Take a deep breath and think about your feelings and the beliefs that fuel them.
3. Put these beliefs to the test and see if they hold water. Be on the lookout for signs that run counter to those notions.
4. Replace those restricting, outmoded beliefs with ones that empower you. Here, affirmations and visualisations may work wonders.

Embracing the Unknown: Overcoming Fear

Having a fear of the unknown, in particular, can be a significant barrier to achieving our goals. Permit me to impart some knowledge regarding the uncertainty principle as it pertains to quantum physics. It's as if one were attempting a simultaneous determination of the exact position and velocity of a particle. There are many unexpected turns in life that might make us feel excited. It's a path full of boundless possibilities begging to be explored.

Let us delve into the fascinating story of J. K. Rowling now. Despite many setbacks, Rowling saw her Harry Potter books and films become a worldwide phenomenon. No amount of anxiety or disappointment could derail her determination. She persisted doggedly, never faltering in her quest to realise her dream. That she overcame her fears and accomplished so much on her journey is quite inspiring.

Strategies for Conquering Anxiety

1. Accept your anxieties about what they are: normal human emotions. Embrace them without judgement. The first step

in conquering adversity is realising what it is and then doing something about it.

2. See your worries as opportunities to learn and improve. Consider the valuable lessons you can learn from their mistakes and triumphs.

3. Loosen up and ride the wave of success as you divide your objectives into manageable chunks. Dealing with it will get simpler and more self-assured as time passes.

4. Get yourself a great support system of people who will cheer you on no matter what and who genuinely appreciate being in your life.

Streamlined Manifestation: A Master Class

It would appear that many people have the wrong idea about manifestation and think it's easy to do. Some people think they can effortlessly bring their desires into existence without really having to do anything by just thinking about them or saying them out. While our ideas and convictions certainly make a difference, it is our deeds that really make our dreams a reality.

Thomas Edison's invention of the lightbulb is a prime example of the power of dogged persistence. Until he achieved his goal, Edison persisted in his pursuit and conducted a great deal of experimentation. His life exemplifies the value of setting goals and working relentlessly to achieve them.

Finding a Middle Ground Between Reflective Thinking and Resolute Action

1. It is essential to have a distinct idea of what you wish to bring into existence in your life. Visualise the outcome you want with pinpoint accuracy.

2. Join forces to figure out a strategy that will help you achieve your goals.

3. Keep striving to accomplish your objectives and maintain your motivation. Doing this will show the universe how dedicated you are and give you momentum.

Keep an open mind and be flexible with your ideas. As you progress through the process of manifestation, you may find that it needs some changes.

Conviction Cultivation: Overcoming Doubt

Being human is all about questioning things, you know? But if we choose to ignore it, it will be harder for us to make our aspirations a reality. Having faith in the procedure is crucial.

When Henry Ford first proposed the idea of mass-producing affordable automobiles, many people were sceptical. The story of Henry Ford shows how unfaltering faith in the face of doubt led to the revolutionary success of the Model T. However, he was steadfast in his belief in his vision and did not waver in his pursuit of it.

The Art of Developing Firm Beliefs

1. Your views and goals can be enhanced by using positive affirmations.
2. Every day, pretend that your dream has already come to fruition.
3. Take a moment to be grateful for all the good things happening in your life. Learn to be grateful for what you have and what you can achieve in the future. When you shift your focus from worrying about shortages to celebrating abundance, you'll find that things become clearer.

Maintain a state of mindfulness and concentrate on the here and now. Keep your cool and have faith that everything will work out as it should.

Dispelling Common Myths Regarding Quantum Mechanics

Understanding and using quantum physics can be a challenging endeavour, frequently resulting in broad misconceptions. Okay, then, I'll clear up some of these misunderstandings:

Many people mistakenly believe that only tiny objects are subject to quantum mechanics.

The scope of quantum mechanics, you know, goes well beyond those tiny subatomic particles. A few important things to think about, if you catch my drift. Entanglement and the observer effect are intriguing ideas because they imply that our consciousness has the power to significantly impact our reality and everything around us.

That Quantum Theory is purely theoretical is widespread yet incorrect.

Beyond its theoretical foundations, quantum mechanics has real-world implications in many domains, such as healthcare and technology. For a little while, let's delve into the intriguing world of quantum mechanics and ponder how MRI machines and semiconductors function on the inside.

Despite its intricacy, quantum mechanics is actually rather easy to understand. It's normal if quantum mechanics' complexities appear overwhelming at first. Breathe deeply and unwind! Anyone may learn the fundamentals of this intriguing subject. Understanding and applying these concepts becomes easy with the help of analogies, simplified explanations, and real-world examples.

Challenging Assumptions

1. Increase your knowledge by looking for materials that break down the complex concepts of quantum mechanics into their parts and present them in an easy-to-understand way. You can learn a lot via books, films, and even online classes.
2. Okay, I'll give you the second one. Maintain an attitude of openness and curiosity as you seek to comprehend the world around you better. If you have questions or would like to learn more, inquire.
3. Try it out and observe the results when you add some quantum principles to your regular regimen. Consider how your ideas and values are shaping your worldview.

Explore the intricate web that connects our sense of identity with our aspirations.

What we go through in life is heavily impacted by how we see ourselves, what we call our self-concept. It is possible to obstruct one's own progress when one's self-perception conflicts with one's goals.

A long time ago, there was a tale that explored the intriguing world of how we see ourselves and how it affects our lives.

Okay, then, let's dive into Oprah Winfrey's fascinating life narrative. Although Oprah had a tough childhood, she overcame adversity by changing her outlook. After years of feeling confined, she broke free and achieved remarkable success by believing in herself and her abilities. Her story exemplifies how changing how we see ourselves may have a dramatic impact on our daily experiences.

Striking a Healthy Balance in Your View of Yourself

1. Pause and think about how you see yourself. Take a deep breath and think about everything going on in your life that could limit you or make you feel unimportant.

2. Got it. The second one is right here. Improving your self-esteem is as simple as making positive affirmations a part of your daily ritual. Statements with the impact of "I deserve abundance" or "I can manifest" can change people's lives for the better.

3. Various approaches to moulding one's own sense of identity

4. Think of yourself as someone who has accomplished all they set out to do. Envision yourself wholly immersed in an alternate reality where you feel and act under every nuance of that reality.

5. Never stray from your authentic self, and always do what makes you happy. Get ready to be amazed by how effortlessly your aspirations will come true.

Having faith and waiting for things to fall into place, getting to know things and believing in them.

Patience and unwavering belief are necessary companions on the path to manifestation. Disrupting the natural flow of energy and impeding growth are impatience and uncertainty.

An enchanted tale unfolded once at a time amid breathtaking natural scenery. This narrative beautifully portrays the power of determination and self-improvement, serving as a powerful reminder of the immense resilience that lives within every individual. I have a fantastic tale to tell you about the bamboo tree.

A profound reflection of manifestation is the growth of the bamboo tree. Growing bamboo from the seed requires both time and care. It only takes a few weeks for this bad boy to build momentum and then skyrocket several feet. It takes a lot of trust and patience to make your dreams come true.

Acquiring the Virtues of Patience and Trust

1. Then why not try some mindfulness exercises? Incorporate deep breathing and meditation into your regular practice. Following these guidelines will help you stay focused and in the here and now.
2. Write your thoughts and feelings about the things that make you happy and the things you want to achieve. You can change your perspective from one of lack to abundance by making this practice a regular part of your life.
3. Let's have faith in the journey. Calm down and quit worrying about details. Always remember that the universe is rooting for your success.
4. Take a moment to appreciate the little wins; they add up to a big difference. These seemingly insignificant occurrences influence your personal growth and strengthen your self-confidence as you progress along your path.

Delving into the Complex Interplay of Ambiguity and Transparency

Inspiring in-depth analysis and deliberate judgement, questioning can be incredibly beneficial. However, we miss out on new possibilities when we let our guard down too much. Finding a happy medium between being sceptical and open-minded helps us to investigate novel concepts without losing touch with reality.

An Original Example: Carl Sagan's Method

The great astronomer Carl Sagan stressed the need to be both sceptical and open-minded. He fostered an environment that was receptive to new ideas and possibilities, while simultaneously encouraging critical thinking and inquiry. A new sense of relatability is generated by Sagan's method, which captures the nuances of equilibrium between scepticism and open-mindedness.

Discovering the Delicate Equilibrium Between Uncertainty and Openness

1. When new knowledge comes your way, take a moment to think about it thoroughly. Inquire and compile proof to back up your claims.
2. Keep an open mind and revel in the wondrous wonders of life. Never close your mind to the possibility of something exciting happening. Seize the opportunity to delve into fresh concepts and gain insightful thoughts from different angles.
3. Strike a balance: Understand that a balanced mentality requires a healthy dose of both openness and scepticism. If we want to discover new things, we need to be curious and ready to challenge our own assumptions.
4. Participate in conversations with people from different backgrounds and experiences so you can learn from their unique points of view. It is a worthy endeavour to explore new ideas and embrace alternative opinions.

It could be quite a voyage to navigate emotional difficulties. The difficulties of life are inevitable, yet life is still a rollercoaster. Realising that we aren't alone in facing emotional challenges is essential while we're dealing with our own, although, at moments, it may appear we are getting engulfed in a hurricane.

Moods like dread, wrath, and depression might obstruct our progress towards our objectives. In order to make room for our deepest dreams, we need to face and manage these emotions. One of the techniques to manage emotions is Emotional Freedom Techniques (EFT).

The Emotional Freedom Technique (EFT), sometimes called tapping, is a novel approach that successfully addresses emotional obstacles by combining cognitive therapy with acupressure. To ease and balance out unpleasant emotions, practitioners gently tap on certain meridian points.

Becoming an Expert at Conquering Emotional Challenges

1. You can learn more about your emotions by just watching them without passing judgement on them. Pay attention to how these feelings and thoughts materialise in your body.
2. EFT tapping unleashes your power to effortlessly let go of negative emotions and conquer any energetic hurdles that may impede your personal growth.
3. One healthy way to get through hard feelings is to keep a notebook and write down your feelings, which could help you unearth patterns and find the profound connections they have.

Take some time out of your day to meditate and develop mindfulness. Find a calm place to sit and let go of any feelings that arise while you observe. Staying grounded and focused is essential.

Investigating How Societal and Ecological Variables Play a Role

People and the environment play an important role in either facilitating or impeding quantum manifestation. Achieving success requires building a positive network and maintaining an encouraging environment.

According to Jim Rohn, our views and actions are influenced by the people we associate with. The people we spend the most time with have a significant impact on our development as individuals. He postulated that we are an average of the five persons we frequently hang out with. If you want to boost your self-esteem and drive, surround yourself with people who are way more successful than you are.

Ways to Handle Environmental and Social Considerations

1. Think about the people in your life and how they affect your outlook and energy levels. Some people draw negativity out of you or make you feel awful to be around them; please avoid or cut ties with them.

2. You can't afford to be around negative influences; instead, surround yourself with positive, encouraging people. Do you ever wish there was a way to connect with others who share your beliefs and aspirations? Meeting new individuals who are enthusiastic and successful will help you in your growth.

3. Environment is equally important. Make sure you're in the best position to achieve your goals by designing your workplace with attention. Purge everything that isn't essential, inject some creativity, and set an upbeat mood.

If you want to keep your energy and concentration levels up, you need to set and stick to boundaries with other people.

Setting Realistic Goals

Focusing too intently on a single result could stifle development and add unnecessary complexity. We can keep our minds and hearts open to the infinite possibilities the cosmos offers if we practice distancing ourselves from our problems.

Many religions and philosophies place a premium on learning to let go or surrender. The importance of letting go of attachments and enjoying the inherent cycle of life is emphasised in Taoism's notion of Wu Wei. This encounter has the potential to bring us immense joy and contentment.

Strategies for Releasing Emotional Bonds

1. Appreciate the Journey: Develop a firm faith in the law of attraction and the skill of drawing your dreams into reality. Accept the reality that things are unfolding precisely as you'd hoped.

2. Don't get caught up in the precise results; instead, concentrate on your intentions. Learn to let go and not get connected to things intensely. Adopt a way of thinking that is receptive to new ideas and is open to learning.

3. Meditation and mindfulness can help you let go of feelings of attachment and live in the now. Perhaps these tips can help you discover a sense of tranquillity and embrace the natural flow of life.

Rather than fixating on the destination, enjoy the ride there as much as possible. Because of this, I have grown as a person, and I am grateful for the experience.

A Deep Dive into the Realm of Quantum Principles and Their Applications

Incorporating quantum ideas into your daily practice is crucial for tackling obstacles and clearing up any misconceptions. Your ideas, beliefs, and deeds may all be easily and smoothly integrated into the process of manifesting your desires through this integration.

By incorporating quantum principles and Neville Goddard's teachings into our daily lives, we can embrace conscious manifestation. Mindful manifesters typically have more success, a more profound insight, and also more meaningful coincidences in their lives.

These are a few simple things you can do every day to help

1. The key to achieving your goals is to set daily intentions that are in line with them. As you move about your day, keep these intentions in mind.

2. Attempt to be grateful every day. Take stock of all the good things happening in your life and the advantages you already have.

3. Imagine yourself doing all that you have set out to do. Visualise the results you want in great detail every day. Let your senses reawaken, and the experience paint a real and vivid image as you sink into it.

4. You may reinforce your empowering ideas and aspirations by making it a habit to recite positive affirmations every day.

The key is to keep yourself well-informed and to act daily in a way that is congruent with your desires. Even modest efforts can have spectacular results.

Wrapping Up: Embracing the Adventure

Manifestation is a life-altering adventure that is absolutely remarkable. Neville Goddard's teachings and the laws of quantum mechanics serve as its theoretical foundation. Together, we can realise our greatest potential and build a world that reflects our most audacious aspirations if we break down barriers and reject preconceptions.

You must approach each stage of your journey with an open mind and a carefree spirit as you embark on your quest. There is an opportunity for growth in every challenge you face, and the knowledge you gain can lead to enlightenment and deep understanding.

Allow yourself to be guided by the enchantment of quantum principles, the strength of conviction, and the precision of purpose. Believe in your inherent ability to shape your own world, which opens doors to limitless possibilities. Every thought, emotion, and choice you make shapes your life, and you have the remarkable capacity to shape your own destiny.

A tranquil quietness enveloped my room as the gentle light of sunset cast the shadows of the towering hills of the Nilgiris through the windows. The gentle rustle of leaves in the wind and the faint sound of a nighttime animal made for an ideal setting for introspection. As the sun began to disappear into the horizon and darkness fell, my mind began conjuring how the ever-present cycle of creation shapes our existence.

Welcome to the Adventure!

The path of manifestation is one of great transformation. Unleashing our full potential is within reach as we press on, challenging our beliefs and exploring our emotions. With open minds and awe in our hearts, we go on this thrilling adventure to discover what the quantum field has in store for us.

The Enigmatic Quantum Realm and the Eternal Dance

The process of manifestation is eternal, like a dance in the quantum field; it never ends. Everything that exists is a product of our mental processes, including our feelings, wants, and ambitions. The key to living life to the fullest is balancing taking part in it all and keeping a sharp eye out for both the opportunities and threats that present themselves.

The Significance of Self-Compassion

As we go through this process, it is essential to be patient and kind to ourselves as we face the difficulties of manifestation. Obstacles and uncertainties are inevitable companions on any path. Try not to be too hard on yourself when you make mistakes; instead, see them as opportunities to learn and develop.

Devoted to the Journey That Lies Ahead

Putting your faith in the process might be difficult, but the payoff of seeing your aspirations come true is worth the effort. Let go of the need to be in charge and accept the natural progression of things. Have faith that everything will work out for the best in due course. A lively bond with life is born out of this faith, which goes well beyond a simple acceptance of the belief that the universe is kind.

Accepting What Is Unknown

Something about the unknown makes people feel a range of emotions, from eagerness to excitement. Totally accepting the mysterious nature of life is embracing the unknown. When faced with the unknown, it's important to have confidence and trust that each step will show you the route forward. When we step into the unknown, we discover a world of limitless opportunities and incredible adventures waiting for us.

A Shared Objective's Influence

The real difference-maker is the realisation of the tremendous power in collective intention. When people band together, their combined actions and energies can make a significant difference, amplifying the effects of change. A group's efforts to accomplish a goal get multiplied when its members work together with a common goal in mind. This is illustrated in many contexts, including social movements, community efforts, and even families striving for their aspirations. Our desires become enchanted when they align with the collective's goals. We are bound by the common goal of achieving extraordinary feats.

A Few Useful Pointers for Regular Training

1. Start your day off right with a simple yet effective morning ritual that will set you up for success. As you get ready for the day, take a few moments to sit back and think. Visualise your heart's desire and let yourself experience the range of emotions that come with it.

2. In order to maintain a focus on your objectives, it is important to allow yourself well-deserved breaks throughout the day. Take advantage of the breaks to strengthen your convictions and picture your aspirations.

3. As you unwind for the night, take stock of your day and be thankful for what you have. Take a deep breath and think about all the new knowledge and improvements you've achieved. Be grateful for everything that is coming together and all the little blessings.

4. Be part of dynamic community and meet others who share your interests. Take part in meaningful discussions that will make you feel accomplished. Become a part of communities or groups that help you achieve your objectives and create an encouraging environment for your personal growth; these are our primary priorities.

Preserving the Magic of Every Moment

While pursuing our goals, it is essential to acknowledge and value even the smallest successes. You can take comfort because you are making progress and heading in the right direction with each milestone. Feel true happiness and appreciation as you relish in your accomplishments. Your will to succeed increases with each victory you taste.

The Effect of Ripples

Beyond your personal reality, manifestation impacts everything in profound and far-reaching ways. Your thoughts, beliefs, and actions impact people around you. Those around you get captivated by your presence and derive inspiration to pursue their aspirations and go on their paths of self-discovery. Many people's perspectives on the world have shifted because of such instances. An increasing number of people are realising their inherent strengths and maximising their potential.

A World with Endless Opportunities

Your path to manifestation is positive proof that the cosmos is a vast arena of possibility. Delve into the teachings of Neville Goddard and the enthralling realm of quantum mechanics, and you will discover new depths of your boundless potential. Every idea, plan, and deed has the tremendous potential to shape the life you desire.

Finally, Some Thoughts

As the sun goes down and the stars come out, lighting up the peaceful hills, it's hard not to wonder how big the universe is. Setting out on the path of manifestation is a lot like exploring the night sky: expansive and full of stars. Every star has untapped potential, like a dormant dream just waiting to be awakened. Our true identities, our highest ambitions, and our limitless potential will be revealed to us as we continue on this path.

Personal development and change are at the heart of the manifestation process. The point is to be ourselves and to accept that we can make a difference in the world. It's all about finding that sweet spot where our souls and the universe harmonise, where our dreams harmonise with the cosmic organic beat. With an attitude of profound appreciation, a desire to explore unexplored regions, and an urge to experience the wonders of life, let's set off on our magnificent adventure.

With confidence that the cosmos is cheering us on, let us set out on this amazing trip with gusto and enthusiasm. Have faith in the journey, celebrate your successes, and have an open mind to the limitless opportunities that lie ahead. Not only will we fulfil our desires on this incredible path of manifestation, but we will also fully embrace and embody our vivid and real selves.

CHAPTER 14

The Art of Quantum Leaping: Becoming Your Desired Reality

It was a brisk evening, and the cool air of the Nilgiris wrapped around me like an old, familiar shawl. The sky above shimmered with stars, each a beacon of infinite possibilities. From the verandah of my little cottage, I watched as the mist rolled over the tea gardens, weaving its way through the hills like a whisper from the universe. My thoughts wandered to the concept of quantum leaping—a practice as profound and mesmerising as the cosmic dance of these distant stars.

Neville Goddard's words drifted into my mind: "Change your conception of yourself, and you will automatically change the world you live in." Quantum leaping isn't a fanciful notion or wishful thinking. It is the deliberate art of stepping into the version of yourself you most desire, bypassing linear time and logic constraints. It invites you to shed the cloak of your old identity and embrace the radiance of your highest self, much like the stars shedding their light on the darkened hills.

What Is Quantum Leaping?

Imagine walking through the vast Nilgiri forests, where every path offers a different destination. Each trail represents a version of your life—a timeline shaped by choices and beliefs. Quantum leaping is the art of choosing a new path, not by shuffling along the old one inch by inch, but by stepping directly into the reality you desire.

Neville often compared this process to 'living in the end'. He believed all possibilities exist simultaneously within the quantum field, much like the stars that scatter the night sky. To quantum leap is to embody the feelings and identity of the version of yourself who has already reached the end goal. It is not about wishing or waiting but assuming the feeling of your wish fulfilled and letting the universe catch up.

For instance, imagine walking into a library with countless shelves containing a different version of your life. Every decision you've ever made—or could have made—exists on these shelves. Quantum leaping is the art of choosing a new book or story and stepping into it. This isn't about incremental change or waiting for time to pass. It's about bypassing linear progression and shifting into a new reality.

In scientific terms, quantum leaping finds its roots in quantum physics, specifically the principles of superposition and the observer effect. These concepts suggest that particles—and, by extension, possibilities—exist in multiple states simultaneously. When you focus on a particular possibility, it collapses into reality.

The Science Behind the Magic

Quantum leaping may feel mystical, but its foundation lies in science. The principles of quantum physics—superposition, the observer effect, and frequency—profoundly mirror Neville's teachings.

1. **Energy, Frequency, and Vibration**
   The Nilgiris hum with energy—be it the rustle of leaves, the gurgling streams, or the distant calls of birds. Everything vibrates at its frequency, including you. Nikola Tesla once said, "If you want to find the secrets of the universe, think in terms of energy, frequency, and vibration." Similarly, Neville believed that aligning your internal frequency—your thoughts, emotions, and beliefs—with your desired reality allows you to step into it.
   Consider Maya, a shy woman who dreamt of becoming a public speaker. For years, she told herself she wasn't cut out for

it. One evening, much like this one, she decided to leap. She began visualising herself confidently, addressing an audience. She felt the applause, the connection, the exhilaration. Maya didn't just think about it; she started embodying it—changing her posture, tone, and habits. Within months, she was living the reality she had once only imagined.

2. **The Observer Effect**

   Quantum physics teaches us that observation shapes outcomes. Neville echoed this with his belief that "your imagination is the instrument, the means, whereby your redemption from slavery, sickness, and poverty is effected." In the same way that observing a particle influences its state, your focused attention on a desired reality brings it into existence.

3. **Parallel Realities**

   Much like the sprawling paths of the Nilgiri forests, every decision you make creates a new timeline. Quantum leaping allows you to consciously align with a version of yourself in a parallel reality.

The Power of Belief and Frequency

Neville said, "You do not attract what you want. You attract what you believe to be true." Belief is the foundation of quantum leaping. Imagine tuning a radio to a station you doubt exists. You cannot access it without the conviction that it is there. Similarly, to jump into your desired reality, you must believe in its existence and align your inner frequency to match it.

Practical Exercise: Embodying Your Desired Reality

1. Sit quietly by a window, perhaps overlooking the serene hills of the Nilgiris.

2. Close your eyes and imagine your desired reality in vivid detail.

3. Feel it as though it is already yours—hear the sounds, see the colours, and sense the emotions.

4. Now, carry those feelings into your present moment. Walk, speak, and act as if you are already living that reality.

Neville called this an 'assumption'. When you assume your desire is fulfilled, you align with its frequency, and the universe mirrors it back to you.

The Role of Inspired Action

Quantum leaping is not passive dreaming. Neville insisted on the necessity of action: "Faith without works is dead." As you shift your frequency, intuitive nudges will guide you. These are the breadcrumbs of your new reality. Follow them.

Aarav, an aspiring entrepreneur from Ooty, exemplifies this. He envisioned himself as a successful business owner, aligning his thoughts and emotions with that reality. One day, he felt compelled to attend a local event. There, he met a mentor who helped him secure his first investment. Aarav's inspired action turned his vision into reality.

Normalising Your Desired Reality

For your desired reality to manifest, it must feel as natural as the air you breathe. Neville called this "living in the feeling of the wish fulfilled."

- **Visualisation:** Visualise daily until your desired reality feels ordinary.
- **Environment:** Surround yourself with people and places that reflect your goals. If you wish to attract abundance, spend time in settings where abundance is evident.
- **Affirmations:** Repeat affirmations like, "I am abundant and successful," reinforcing your belief in your new reality.

Overcoming the In-Between Phase

The journey between your current reality and the one you desire can feel like crossing a river. You've let go of one shore but haven't yet reached the other. Neville's advice is simple: persist. He would say, "Don't condition your desire. Trust it is already done."

The Ladder Approach to Belief

Sometimes, the leap to your ultimate goal may feel too vast. Neville suggested starting with smaller assumptions to build faith.

For instance, if your goal is to earn ₹5,00,000 a month but you currently earn ₹50,000, your first assumption could be ₹1,00,000. As you normalise this, you can leap further.

Practical Steps to Quantum Leap

1. **Clarify Your Desire:** Define your goal with precision and detail.
2. **Visualise Daily:** Imagine your desired reality vividly, infusing it with emotion.
3. **Shift Your Beliefs:** Identify limiting beliefs and replace them with empowering ones.
4. **Act As If:** Embody your desired self's habits, feelings, and actions.
5. **Follow Intuition:** Trust and act on inspired ideas.
6. **Detach:** Release the need to control the outcome and trust the process.

Conclusion: The Quantum Creator Within

As the stars above the Nilgiris twinkle like promises waiting to be fulfilled, I am reminded of Neville's timeless wisdom: "Man moves in a world that is nothing more or less than his consciousness objectified."

You are the author of your story, the sculptor of your reality. Each thought, each emotion, and each action is a brushstroke on the canvas of your life. So, what masterpiece will you create? What reality will you choose to leap into?

The universe is listening. Leap and let its magic unfold.

CHAPTER 15

Becoming Nobody: A Journey into the Zero-Point Field

The Nilgiris stretched before me, veiled in a soft mist that blurred the line between earth and sky. From my perch by the window, the world appeared still, yet beneath the stillness lay an unseen pulse—a hum of life connecting every leaf, every breeze, and every ripple in the stream. It was the zero-point field, the invisible fabric that holds the universe together. Neville Goddard's teachings came to mind: "The world is yourself pushed out." I saw it clearly in this stillness—the universe, and I were not separate but part of the same eternal field of energy and potential.

The zero-point field, a cornerstone of quantum physics, is not just a scientific theory. It is the fertile void, the birthplace of all creation, and the space where infinite possibilities converge. To access it, one must go beyond the known, beyond identity, and into the boundless unknown. In the act of becoming nobody, it is here that the true journey of creation begins.

The Zero-Point Field: The Source of All Creation

Neville often spoke of 'entering the silence', a state where the mind is stilled and the self dissolves. This silence mirrors the zero-point field, the infinite void from which all things arise. Imagine the moment before the Big Bang—a vast stillness brimming with potential, waiting to burst into existence. This is the zero-point field, the essence of creation.

This field exists beyond time and space, a realm of pure energy and infinite possibilities. Everything you see—the verdant tea gardens of the Nilgiris, the towering eucalyptus trees, the sun breaking through the mist—originates from this field. As Neville taught, "You are not separated from the thing you desire." In the zero-point field, there is no separation. Everything you long for already exists in potential form, waiting for you to bring it into reality.

The Descent into Density

As beings of energy, we descend from this universal intelligence into the density of the physical world. Here, we define ourselves by our roles, possessions, and circumstances. "I am a teacher," we say, or "I am wealthy," or even "I am struggling." These labels form the scaffolding of our identity, but as Neville often warned, "You must abandon yourself to your ideal as if it were true."

I recall a conversation with a friend, Priya, who longed to start her own business but felt trapped by fear of failure. Her thoughts were anchored in her current reality of doubt and lack. "I can't see it happening," she admitted. Her fear and self-doubt were like weights, keeping her tethered to a lower frequency. To manifest her dream, Priya needed to release her attachment to her current identity and step into the unknown, where her vision already existed in the zero-point field.

The Problem with Lower Frequencies

Every thought emits a frequency, much like a radio signal. We operate at lower frequencies when we dwell in fear, resentment, or doubt, slowing down the manifestation process. Neville described this as "living in the shadow of your self-imposed limitations."

I once met a man stuck in a financial struggle cycle. His thoughts were dominated by phrases like, "I'll never get out of this," or "Why does this always happen to me?" These low-frequency beliefs perpetuated his reality. It wasn't until he shifted his focus—cultivating gratitude and imagining abundance—that his circumstances changed. Neville would have called this "changing your conception of yourself."

The Path to the Unified Field: Becoming Nobody

To access the zero-point field, we must let go of everything we think we know about ourselves. This act of becoming nobody is not an abandonment but a liberation—a return to the boundless potential within.

1. **Letting Go of the Known**

   Neville often spoke of the 'silence of the senses'. Imagine sitting quietly, eyes closed, as you release all attachments to your body, possessions, and roles. In this stillness, you transcend the physical and enter the 'generous present moment', a timeless space where creation begins.

   I once guided a corporate executive through this process. At first, she resisted, clinging to her identity as a high achiever. But as she persisted, she described a profound peace as though she had stepped into an ocean of calm. At that moment, she connected with the zero-point field.

2. **Cultivating Elevated Emotions**

   Neville taught that feelings are the secret to creation. To align with the zero-point field, we must cultivate emotions like gratitude, love, and joy, which elevate our frequency and bring us closer to infinite possibilities.

The Science of Oneness

Einstein's equation reminds us that matter and energy are interchangeable. Beyond the speed of light, all matter dissolves into pure energy, uniting everything in a state of oneness. Neville described this as "living in the end." When we imagine ourselves as one with our desire, we dissolve the illusion of separation and align with the quantum field.

Brain Coherence and Heart Intelligence

The brain and heart are gateways to the zero-point field. When thoughts and emotions harmonise, we emit a powerful electromagnetic signature that aligns with the quantum field.

- **Brain Coherence:** Meditation shifts the brain from stress-induced beta waves to the relaxed, creative alpha and theta states. This quiets the analytical mind and opens the door to the subconscious, where our deepest beliefs and desires reside.
- **Heart Coherence:** The heart generates an electromagnetic field far more potent than the brain. By cultivating gratitude and love, we amplify this field, deepening our connection to the zero-point field. Neville described this state as "feeling the wish fulfilled."

Practical Steps to Access the Zero-Point Field

1. **Meditation:** Dedicate time daily to stillness. Imagine becoming a nobody, releasing all attachments and identities.
2. **Elevated Emotions:** Practise gratitude and love, which raise your vibration and align you with infinite possibilities.
3. **Visualisation:** Imagine your desired reality as if it already exists. Feel it, see it, and live it in your mind.
4. **Surrender to the Unknown:** Trust the process. Let go of control and allow the universe to realise your vision.

The Beauty of Becoming Nobody

As the sun dipped below the horizon, the mist over the Nilgiris thickened, enveloping the hills in a tranquil embrace. In that stillness, I realised that the journey to the zero-point field is not about becoming more but about becoming less—less constrained, less fearful, and less attached.

When we surrender to the infinite unknown, we step into the role of co-creators with the universe. We shape reality not through force but through alignment, discovering that we have the limitless potential to become everything by becoming nobody.

This is the essence of quantum creation: to dissolve into the infinite so we can shape the finite, to quiet the self so we can hear the universe, and to become nobody to create everything.

CHAPTER 16

The Power of Silence: Unlocking Quantum Creation Through Stillness

The timeless and serene Nilgiris lay shrouded in a soft veil of mist. The eucalyptus groves swayed gently in the mountain breeze, their whispers carrying the scent of earth and pine. Silence reigned over the hills—not an absence of sound, but a living stillness, a force that seemed to cradle the world in its quiet power. In this sacred hush, the wisdom of Neville Goddard stirred, resonating with the mysteries of quantum physics. Deep and profound silence mirrored Neville's teachings: stillness is the cradle of creation. Within this stillness, the infinite potential of the quantum field unfolds, waiting for the conscious observer to mould it into form.

The Creative Power of Stillness

"Be still and know that I am God." Central to Neville's philosophy, these words speak of the divine power that lies dormant in silence. In stillness, the imaginative faculty awakens—the creative force through which desires are realised.

In the realm of quantum physics, this parallels the concept of the observer effect: particles remain in a state of infinite potential until observed, collapsing into a definite reality. Neville taught that our imagination is the observer, the force that collapses possibilities into manifestation. By dwelling in the silence of fulfilled desire, we bridge the gap between what is and what could be.

Nature's Lesson: The Gestation of Desire

The Nilgiris, with their rolling tea gardens and mist-cloaked valleys, embody the rhythm of silent creation. The tea bushes thrive not through force but through the quiet cycles of sun, rain, and soil. Likewise, Neville often spoke of desires as seeds planted in the fertile soil of consciousness. They must be nurtured with patience and trust, allowing them to take root and grow silently.

Neville warned that speaking prematurely of one's desires dissipates creative force, like exposing a seed to harsh elements before it sprouts. In quantum terms, this reflects the collapse of the wave function: when a possibility is disturbed by external observation, its coherence weakens. The dreams we hold close, like the roots of ancient Nilgiri oaks, thrive best in the hidden sanctuaries of our minds.

Insulating Consciousness: The Ark of Assumption

The vastness of the Nilgiris remains untouched by modernity's discord, offering a sanctuary for reflection and growth. Similarly, Neville **emphasised** the importance of insulating one's consciousness from external doubts and limiting beliefs. He called this the "Ark of Assumption," a protective vessel that preserves the purity of one's inner state.

Quantum physics offers a striking analogy: systems in resonance amplify their vibrations, while discordant frequencies disrupt coherence. Surrounding ourselves with those who resonate with our old fears and limitations weakens the vibrational alignment necessary for manifestation. Silence, then, becomes a shield, amplifying the resonance of our desires.

The Three Chambers of Silence

Neville's teachings on silence transcend mere quietude; they guide us through three progressive chambers of stillness:

1. **Physical Silence:** Refraining from sharing desires with others.
2. **Mental Silence:** Silencing the internal dialogue of doubt and fear.

3. **Spiritual Silence:** Living as though the wish has already been fulfilled.

These chambers echo the quantum state of coherence, where particles align in perfect harmony. In this state of alignment, manifestation becomes not an effort but an inevitability.

Resonance and Transformation

Seasons in the Nilgiris transition seamlessly, reshaping the landscape with each passing phase. This natural rhythm reflects Neville's principle that life mirrors our inner state. When we change our state of being, our world rearranges to match it.

Quantum superposition reinforces this truth: particles cannot exist simultaneously in two opposing states. Likewise, we must fully commit to a new consciousness, releasing the old. Neville's insight was clear— when you assume the feeling of the wish fulfilled, your entire reality reflects that assumption.

The Cocoon of Sacred Isolation

The metamorphosis of a caterpillar into a butterfly offers a powerful analogy. Within the silence of its cocoon, the caterpillar undergoes a profound transformation, emerging as something entirely new. Neville referred to this process as the "Law of Silent Transformation."

Quantum physicists describe this as zero-point energy—a vacuum with infinite potential. In this sacred isolation, the old self dissolves, and the new self emerges fully aligned with its desires. Silence, therefore, is the cocoon in which our transformation unfolds.

The Seven Pillars of Silence

Outlined below are seven principles to protect the creative process:

1. **Physical Withdrawal:** Avoid environments that reinforce the old state.
2. **Social Disconnection:** Distance from those tied to limiting beliefs.
3. **Conversational Silence:** Refrain from discussing your desires.

4. **Mental Quietude:** Still the voice of doubt within.
5. **Emotional Stillness:** Cultivate unwavering confidence in your vision.
6. **Spiritual Insulation:** Guard your consciousness from disbelief.
7. **Complete Absorption:** Immerse **entirely** in the feeling of the wish fulfilled.

Breaking even one pillar risks weakening the alignment between your inner state and the desired reality. Silence fortifies these pillars, ensuring the manifestation process remains intact.

Emergence and Manifestation

When the morning mist lifts in the Nilgiris, the landscape reveals itself anew, bathed in light and clarity. This is Neville's "Law of Silent Emergence": manifestation occurs naturally, without force, when the gestation of silence is complete.

Quantum physics echoes this through spontaneous symmetry breaking—a system in equilibrium shifts into a new, stable state. Transformation, like dawn breaking over the Nilgiris, is subtle and profound.

The Symphony of Silence

As dusk settles over the Nilgiris, the mountains hum with an ancient stillness. Streams murmur, stars shimmer above, and the wind whispers of eternity. This symphony of silence is where creation begins and ends.

Neville Goddard and quantum physics converge on this timeless truth: within the stillness of your consciousness lies the power to reshape your world. By embracing sacred silence and aligning with the quantum field of infinite possibilities, you become the architect of your reality.

The Nilgiris, in their quiet majesty, remind us of this eternal wisdom: silence is not emptiness but the fullness of all that is yet to come.

CHAPTER 17

Energy Management: Harnessing Quantum Power in the Blue Mountains

The Nilgiris, known as the Blue Mountains, cradle an ancient wisdom that hums through their mist-cloaked hills. As the morning light filters through towering eucalyptus trees and the koel's song fills the air, the mountains seem alive with an energy that resonates with the rhythm of life itself. It's here, amidst the timeless beauty of this landscape, that the principles of quantum creation and energy management come to life.

Sitting on the porch of a rustic wooden cottage, I sip my cardamom-laced chai and watch the sun cast its golden glow over the valley. The crisp air clears your mind and invites you to explore the deeper truths of existence. This morning, the forest whispers of something profound—how we can harness our energy to shape our thoughts and reality.

The Dance of Energy and Awareness

Energy is not a mystical concept reserved for esoteric teachings; it is a scientific truth that forms the foundation of our existence. Quantum physics reveals that every atom, particle, and thought vibrates with energy. This energy connects us to everything around us—like the unseen currents of wind weaving through the shola forests.

The first step to mastering this energy is awareness. Close your eyes and focus on your breath. Feel the air fill your lungs and flow out. Notice the subtle tingling in your hands and feet. That is your energy, your life force, always present but often overlooked. This simple awareness opens a deeper connection with yourself and the universe.

Padmini's Journey: An Awakening

Padmini, a friend burdened by the chaos of city life, came to the Nilgiris seeking solace. "I feel disconnected," she admitted one evening as the valley bathed in twilight hues. I guided her to start with something simple yet profound—breathwork.

"Close your eyes," I said, "and take a deep breath. Feel it nourish you. Let your mind quiet, and your body speak."

At first, Padmini was sceptical, but as the days unfolded, she noticed subtle shifts. The forest seemed alive, her senses sharper, and her emotions more balanced. By the end of her stay, she discovered how to channel her energy toward healing. Her recurring headaches disappeared, and she described the experience as finding a compass within herself.

The Power of Breath

Breath is the bridge between our conscious and subconscious minds. It connects us to our body's energy and provides a powerful tool for resetting our mental state. Deep, rhythmic breathing activates the parasympathetic nervous system, signalling your body to relax and conserve energy.

Here in the Nilgiris, mindful breathing feels natural. Try this: inhale deeply through your nose, hold for a moment, and exhale slowly through your mouth. Visualise the breath, carrying away tension and filling you with vitality. This practice grounds you, aligning your energy with the steady pulse of nature.

Shaping Reality with Visualization

One misty morning, I invited Padmini to practice quantum visualisation. "Close your eyes," I said. "Picture your goal—not as a distant dream but as a present reality. See it, feel it, and believe it."

Padmini imagined herself confidently leading a team, her voice clear and her presence magnetic. She described the details vividly— the warm light in the room, the attentive faces, and the applause. Weeks later, she called to share an uncanny coincidence: a leadership role that mirrored her visualisation falling into her lap.

Quantum physics supports this phenomenon. Vivid mental images activate the same neural pathways as physical experiences. When you combine visualisation with emotion—gratitude, joy, or excitement— you amplify your signal to the quantum field, aligning it with your desired reality.

Scripting Your Future

Scripting is another powerful tool for directing energy. Think of it as writing your life story, but in the present tense, as if your dreams have already come true. Padmini embraced this practice during her stay. She wrote in her journal each morning: "I am thriving in my dream role, surrounded by supportive colleagues. I feel confident, fulfilled, and aligned with my purpose."

This act of scripting engages your subconscious mind, reinforcing your intentions and drawing opportunities into your life. It's a practice that turns thoughts into reality, one written word at a time.

Transforming Emotional Energy

The Nilgiris taught Padmini a vital lesson: emotions are not obstacles but energy in motion. Anger, fear, and sadness carry a potent charge that can drain or fuel you, depending on how you channel it.

When frustration clouded her mind one afternoon, I suggested she walk it off. As she climbed the hill, each step transformed her anger into determination. By the time she reached the summit, she felt a renewed clarity. Emotions are akin to the storms that ravage the forest, which we can direct and harness to bring growth and strength.

Acceptance is the key. Fighting emotions wastes energy; embracing them allows you to redirect their power constructively.

Breaking Patterns and Building New Pathways

Neuroplasticity—the brain's ability to rewire itself—mirrors the regenerative cycles of the Nilgiris. As the forest grows new paths after a storm, we can create new neural pathways through consistent effort.

Start small. Link new habits to existing routines. For example, if you want to cultivate gratitude, practice it after brushing your teeth. These habit anchors create triggers that make change easier. Over time, the new behaviour becomes second nature, much like a well-trodden trail in the forest.

Breaking old patterns is equally important. Try something different—a new route to work, a fresh perspective on a problem. Each slight shift disrupts the automatic loops that keep you stuck and opens the door to new possibilities.

Synchronicity: Signs of Alignment

As Padmini practised these tools, she began noticing synchronicities—meaningful coincidences that aligned with her intentions. A chance encounter led to a valuable connection, and an unexpected email offered a new opportunity. These weren't random events; they were the universe responding to her focused energy.

Pay attention to such moments as signs that you're tuned into the quantum web, where your thoughts and reality converge.

Neville Goddard's Insights

As the mist lifted from the Nilgiris, revealing sunlit valleys, I thought of Neville Goddard's timeless teachings. Neville emphasised the power of imagination and the feeling of the wish fulfilled. His principles align seamlessly with the quantum tools we've explored.

Neville taught that your inner state creates your outer reality. By embodying the emotions and beliefs of your desired state, you draw them into your life. His insights remind us that the subconscious mind, fueled by repetition and faith, is the key to transformation.

Conclusion: The Energy Within

The Nilgiris, with their boundless energy and timeless wisdom, remind us of our infinite potential. Every breath, every thought, and every choice sends ripples through the quantum field, shaping our reality.

Take a deep breath and feel the power within you. The forest whispers its eternal truth: you are a creator, connected to the universe in ways more profound than you can imagine. Harness your energy, align your intentions, and watch as your life transforms—one conscious breath, deliberate thought, and inspired action at a time.

CHAPTER 18

The Quantum Symphony: Tuning into the Universe's Harmonics

In the infinite dance of the universe, there is a melody—a silent symphony of vibrations, light, and sound. This is the quantum field: a boundless ocean where frequencies and plasma weave together to form the reality we experience. As I delve deeper into the mysteries of this quantum landscape, I find that everything—every emotion, thought, and action—is part of an incredible harmonic structure.

Neville Goddard often described this 'melody' as the interplay of consciousness and imagination. He taught that our imagination is not separate from reality but the force that creates it. The symphony we perceive externally originates within us, orchestrated by the frequencies of our inner state.

The Plasma Principle: Building the Fabric of Reality

Plasma, often described as the fourth state of matter, is more than just the fiery glow we associate with stars. It is the fundamental structure of creation itself. Imagine the universe as an intricate tapestry, with plasma threads—composed of light and sound frequencies—binding everything together. These threads vibrate, interact, and form patterns that serve as the building blocks of reality.

Neville Goddard's insight resonates here: "Imagination creates reality." Plasma mirrors our thoughts and emotions as the building blocks of the universe. As two plasma frequencies resonate to create a third, so do our thoughts, feelings, and beliefs harmonise to manifest a tangible reality. Goddard's teachings remind us that this process is not external; the quantum symphony plays within us, directed by the power of focused imagination.

The Black Hole Gateway: Portals to New Realities

Black holes, traditionally seen as voids that consume everything in their path, hold a different meaning in the quantum realm. Here, they are portals—gateways facilitating transitions from reality to reality. These transitions aren't physical journeys but shifts in consciousness and vibration.

Neville described these 'gateways' as moments of assumption. When we assume the feeling of our wish fulfilled, we shift realities, aligning with the version of ourselves already living that reality. Just as a black hole bends time and space, assuming bends the rules of the physical world, collapsing time to bring desires into the present.

Manifestation as Quantum Artistry

Manifestation is an act of co-creation, where intention interacts with the quantum field to produce tangible results.

Harmonise Your Inner World:

Start by aligning your inner frequency with gratitude, joy, and love. These high vibrational states create a fertile environment for manifestation.

- Neville would call this "living in the end"—feeling gratitude and joy as if your desire is already fulfilled.

Imprint the Quantum Matrix:

Visualise your desired outcome with precision and emotional clarity. This creates a geometric pattern within the quantum plasma, directing energy toward your intention.

- Neville's technique of 'revision' aligns with this. You reprogram the quantum field and shift your reality by rewriting a past event in your mind with a favourable outcome.

Release Expectations:

One of the most profound lessons I've learned is to let go of how outcomes should unfold. When we impose expectations, we limit the quantum field's potential to deliver results in unexpected and extraordinary ways.

- As Neville said, "You are not to influence others. Assume your desire is fulfilled and let the infinite intelligence of the universe bring it to pass."

Quantum Harmonies in Relationships

Our relationships are projections of the quantum field, shaped by the frequencies we emit. These qualities are mirrored back to us if we radiate kindness and understanding. Conversely, unresolved inner conflicts often manifest as external discord.

Neville taught that "Everyone is you pushed out." This means that the dynamics we experience in relationships reflect our inner state. To transform a relationship, assume the feeling of harmony and connection. Imagine the other person responding with love and understanding, and watch as your inner alignment creates shifts in the external world.

The Heart: Humanity's Quantum Tuning Fork

The heart is the most potent instrument in this symphony. Unlike the mind, which operates through logic and analysis, the heart resonates at frequencies of unconditional love and unity. This resonance is the key to harmonising with the quantum field.

Neville emphasised the importance of feeling: "Feeling is the secret." By engaging the heart in our manifestations, we access the deepest levels of the quantum field. Practices like heart-centred breathing amplify this connection, allowing the heart to act as a tuning fork that aligns us with our desires.

The Infinite Symphony

Understanding the quantum field transforms life into a deliberate act of creation. Each thought, emotion, and action is a note in the grand symphony of existence. When we align with the harmony of the universe, we gain the power to compose a reality that reflects our highest potential.

Neville Goddard's insights teach us that this symphony begins within. We become co-creators in this infinite dance by imagining the life we desire, feeling it as true, and trusting the process. The quantum field is always listening and responding. Tuning our consciousness to the right frequency allows us to move through realities with grace, intention, and love.

CHAPTER 19

The Law of Trinity: Gratitude, Reset, and the Geometry of Manifestation

Sitting by my window in the serene Nilgiris, surrounded by the rolling blue hills of the Western Ghats, I find myself immersed in the rhythm of life. The gentle rustle of eucalyptus leaves, the occasional call of a Malabar whistling thrush, and the distant hum of a tea factory weaving its melody through the cool air—all these elements seem to whisper a timeless truth. Like these hills, life is a harmony of patterns, connections, and balance.

One such principle that governs the fabric of reality is the Law of the Trinity. It is a cosmic symphony where intention, action, and alignment merge to create new realities. As I sit here, reflecting on the mysteries of existence, I am reminded of how this law can transform the mundane into the miraculous.

Understanding the Law of the Trinity

The Law of the Trinity is not as abstract as it sounds. It represents the synthesis of dualities—positive and negative, action and reaction, thought and manifestation. When two forces interact in resonance, they birth a third, more refined reality. This third state is the culmination of opposites, the 'child' of their union.

Neville Goddard's insights often echo this principle. He spoke of the 'bridge of incident', where thought and feeling merge to bring desires into form. Imagine a village marketplace here in the Nilgiris: two traders exchange their goods—one offers fresh spices, and the other offers honey. Their interaction creates a third reality, a meal enriched by the union of these flavours.

In the realm of quantum creation, the Law of the Trinity is a framework for navigating parallel realities. When you hold a vision, align your emotions with gratitude, and act sincerely, the universe weaves these energies into a coherent reality. Gratitude, as Neville often emphasised, is not just a polite sentiment; it is the force that amplifies and stabilises the vibrations of creation.

Gratitude: The Magnetic Field of Creation

It was Lakshmi, a tea plucker I often met on my walks through the plantations, who taught me the quiet power of gratitude. Every morning, as she worked under the rising sun, she would hum a soft prayer of thanks—not just for the tea leaves she harvested but for the rains that nurtured the plants and the earth's warmth beneath her feet.

Neville described gratitude as "accepting the gift before receiving it." I saw this profound principle in Lakshmi's simple act of giving thanks. Gratitude acts as a magnetic field, drawing abundance not by demand but by resonating with it. Scientifically, gratitude raises the frequency of your thoughts, aligning them with the desired outcomes. By being grateful for the 'unseen', you bridge the gap between your current state and the reality you wish to manifest.

Reset: The Power of Starting Anew

There are days when the rhythm falters—when the melody of life slips into discord. I think of Rajan, a local craftsman who carves exquisite wooden figurines on such days. One day, I found him gazing at a half-finished piece with frustration.

"The reset," he told me, "is the most important part of the process. It's when you step back, breathe, and start again with clarity."

Neville would have called this 'revision'. He taught that by mentally rewriting an experience as we wish it to be, we dissolve resistance and align with the desired outcome. The reset is not about abandoning your goals but realigning with your core. It is when you stop fighting the storm and instead let it pass, trusting the sun will shine again.

To reset is to embrace the art of letting go. Like Rajan stepping away from his carving to regain perspective, resetting allows us to recalibrate our vibration and move forward with renewed purpose.

The Geometry of Manifestation

Picture the Nilgiris themselves: a majestic triangle of land rising from the plains, its three peaks forming a stable foundation. This triangle mirrors the geometry of manifestation—intention, emotion, and action. Without one, the structure collapses, and the manifestation falters.

1. **Intention: What do you truly desire?**
   Your intention must be clear, specific, and free from ambiguity. It is the cornerstone of your creation. Without intention, the universe has no roadmap to follow.
   - Neville often said, "You must know exactly what you want before you can assume it is true."

2. **Emotion: Do you feel the joy of already having it?**
   Emotion is the vibration that signals the quantum field of what to construct. A heart filled with gratitude, joy, or excitement resonates at a frequency that aligns with your desires.
   - "Feeling is the secret," Neville would remind us. Emotion is the energy that brings intention to life.

3. **Action: Are you taking steps in alignment with your goals?**
   Action is the physical expression of your intention. Without action, intention and emotion remain ideas, unable to materialise.
   - Neville described the action as the natural movement that arises from belief. When you genuinely believe, inspired actions follow effortlessly.

Living the Trinity

One misty evening, I met a young boy named Arjun who dreamt of becoming a wildlife photographer. He would wander the forest trails with a borrowed camera, capturing fleeting moments of beauty—the flick of a leopard's tail or the silhouette of a bison at dusk.

But Arjun doubted his abilities, convinced he couldn't succeed without formal training.

"What if," I suggested, "you keep your best photograph pinned where you can see it daily?"

He did just that. Over time, his confidence grew as he saw his collection of photos expand, each one more striking than the last. Arjun had unknowingly aligned with the Law of the Trinity. His intention was clear—to capture the essence of the Nilgiris through his lens. His gratitude for the beauty he witnessed infused his work. And his action—practising daily—brought his dream closer to reality.

Neville would have called this "living in the end." Arjun began to feel the success of his journey long before it manifested fully.

The Whisper of the Hills

As I sit here, pen in hand, the Nilgiris whisper their timeless truth: life is a series of choices, moments where we can create, destroy, or reset. The Law of the Trinity reminds us that creation is a collaboration between our inner world and the universe around us. Gratitude steadies the hand that draws the blueprint, while action carves it into reality.

And so, I leave you with this: Be intentional in your desires, joyful in your gratitude, and unwavering in your actions. When the storms come—and they will—reset with grace. Like these hills, life is not static but ever-changing, ever-beautiful, and ever-ready to align with your most authentic self.

CHAPTER 20

The Alchemy of Content Creation: Manifesting Influence and Authenticity

In the heart of the Nilgiris, under the cool embrace of eucalyptus trees, I sat with my notebook and pen, pondering the essence of storytelling. The rolling blue hills around me stretched into the misty horizon, their tea gardens shimmering like emerald quilts. The distant hum of a train winding its way up the mountain and the soft chirping of cicadas seemed to weave together nature's narrative.

And there, amidst this quiet wisdom, I realised that content creation—whether written, spoken, or visual—follows the same principles as the alchemy of these hills. It transforms thoughts into stories that resonate, influence, and endure.

Neville Goddard's insights whisper through this realisation. "You are the operant power," he taught, reminding us that the world we create mirrors our inner state through words, actions, or presence. Like life, content creation is not an act of chance but a deliberate manifestation of authenticity and intention.

The Quantum Nature of Content Creation

Content creation, like the quantum field, operates in layers of vibration and frequency. At its core, it projects one's energy into the digital or physical world. Every post, podcast, or video manifests one's inner state, shaping the realities of those who consume it.

Imagine a drop of ink falling into a pool of water. The ripples spread outward, colouring everything they touch. This is content in motion—an idea rippling through minds and hearts. But its resonance depends on its origin. Does it emerge from ego-seeking validation? Or does it spring from authenticity, aiming to serve and connect? Neville often said, "An assumption, though false, if persisted in, will harden into fact." Authentic content, born of sincerity, resonates on higher frequencies, creating waves of transformation.

The Heart of Giving in Content

Compelling content begins with the energy of giving. It is the joy of offering value without strings attached—a pure act of creation. Neville believed that actual creation comes from the assumption of abundance. When we give without expectation, we operate from fullness, drawing more into our lives.

Take the example of Malathi, a teacher from Coonoor. She started a YouTube channel to teach spoken English to rural students. She didn't charge a fee or set conditions; instead, she offered her lessons freely, guided by a genuine desire to uplift others. Within months, her channel grew organically. Students began sharing and translating her lessons into other languages, multiplying her impact beyond her imagination. Her success wasn't calculated—it was a natural manifestation of her giving spirit.

The Alchemy of Authenticity

Authenticity is the golden thread in the tapestry of content creation. In a world flooded with influencers, what sets someone apart is their ability to connect with people at a deeply human level. Neville Goddard often spoke of 'living in the end', embodying the feeling of your wish fulfilled. Authenticity extends this concept—living as your most authentic self and creating from that place.

Consider Vidya Vox, a music artist blending Indian and Western styles. Her unique voice emerged not from mimicking trends but from her authentic experience as a cross-cultural individual. Her content resonates because it carries the essence of her journey.

Closer to home in the Nilgiris, there's Shankar, a photographer who captures the everyday lives of tea workers. His photos tell stories of resilience and simplicity. Shankar doesn't chase trends or algorithms; he creates from a place of connection, allowing his work to inspire and resonate naturally.

The Structure of Influence

To create content that genuinely influences, one must understand the psychology of engagement. The human brain processes information best in short, impactful bursts. Neville's idea of focused imagination aligns perfectly here: a single vivid scene imagined with clarity carries more power than hours of scattered thought.

The Nilgiris themselves offer a metaphor. A steady stream carves through mountains over centuries, but the cascading waterfall captures attention and awe. Similarly, your content must balance depth with brevity, offering value in every interaction.

Gratitude and Collaboration: The Keys to Longevity

Gratitude amplifies abundance. Acknowledging your audience, collaborators, and critics creates a feedback loop that sustains your work. Neville's teachings on gratitude are clear: to feel grateful for what you desire as though it is already yours is to align yourself with its manifestation.

Consider an Ooty baker whose Instagram account highlights her recipes, the farmers who grow her ingredients, and the customers who support her business. This act of acknowledgement has turned her small bakery into a thriving hub. Gratitude weaves connections that ripple outward, drawing abundance in unexpected ways.

Collaboration, too, multiplies impact. When creators join forces, they amplify each other's strengths. Take the example of a collective of Nilgiri artists who pooled their talents to host creative retreats. Their combined energy created experiences that drew participants from across the country, turning their small town into a centre of inspiration.

Reset: Adapting and Evolving

Content creation is not static; it demands constant resetting and evolving. As the Nilgiris transform with the seasons, creators must adapt to new audiences, technologies, and trends. Resetting doesn't mean abandoning your essence—it means reinterpreting it to stay relevant.

Neville's concept of 'revision' teaches us that we can reshape our narrative at any moment. Whether it's refining a failed idea or shifting direction entirely, the power lies in our hands.

When I first began writing stories, they were long, winding tales penned for my satisfaction. But as years passed, I learned to condense my narratives for younger readers without losing their souls. Resetting is not a loss—it is a creative renewal.

Living the Creator's Journey

Content creation is a journey of growth, connection, and discovery. It blends quantum science, human psychology, and heartfelt storytelling. Every piece of content—like each leaf in the Nilgiri tea gardens—carries the potential to brew warmth, connection, and transformation.

Neville Goddard reminds us that creation is not about force but faith. "Believe it in, and it will come to pass," he said. As creators, our role is to trust the alchemy of our authentic expressions.

So, step into this journey with courage and sincerity. Let your work reflect the highest vibrations of your heart. Trust that the universe, like the hills of the Nilgiris, will amplify your efforts, shaping them into something enduring and beautiful.

CHAPTER 21

The Quantum Art of Neutrality: Crafting Realities Beyond Perception

In the heart of the Nilgiris, where the mist rolls over emerald tea gardens, and the aroma of eucalyptus fills the crisp mountain air, I sat by a quiet brook pondering the vastness of the quantum field. The chirping of bulbuls blended seamlessly with the rustle of leaves, and it struck me— life is an infinite interplay of creation, observation, and recalibration. Like these hills, the quantum field offers endless possibilities, waiting for us to shape them with our consciousness.

Neville Goddard's teachings came to mind: "You are the creator of your reality." He spoke of the power of imagination, not as an escape but as the foundational tool for crafting the realities we experience. This chapter explores the uncharted realms of neutrality, fractal time, and merging parallel realities, illuminating how we master our holographic existence.

Zero-Point Field and Neutrality: The Interplay of Infinite Creation

In the Chapter 15, we discussed the zero- point field. It's interesting to see how these two concepts contrast but are interrelated as well.

In the quiet vastness of the Nilgiris, where time seems to dissolve into the misty horizons, the concepts of neutrality and the zero-point field emerge as profound truths. Neutrality, like the stillness of the tea gardens at dawn, is the calm within—the art of detachment, where judgment and resistance fall away, and all that remains is an open space for creation. It is the silence before the first note of a melody, the unspoken breath before a word takes form. In this state, we no longer strive or resist; instead, we align ourselves with the gentle flow of life.

On the other hand, the zero-point field is the unseen ocean beneath that stillness—a field of infinite energy and potential where all possibilities reside. It is the blank canvas upon which the universe paints its masterpieces. Neutrality is the doorway to this field. When we release our emotional turbulence, quieting the mind and heart, we step into this limitless expanse where our thoughts and intentions begin to shape reality.

What makes these two concepts inseparable is their harmony. Neutrality allows us to surrender control and enter the zero-point field, while the zero-point field amplifies the clarity and purity of our intentions. Together, they form the foundation of quantum creation. Like standing at the edge of a tranquil lake in the Nilgiris, the reflection of your desires begins to form only when the water is still. And in that stillness, the infinite becomes tangible, weaving dreams into reality.

This interplay is not just a theory—it is a rhythm we live by, as timeless as the hills. Neutrality grounds us, and the zero-point field lifts us into infinite possibility, reminding us that creation is a quiet art and an endless dance.

Neutrality: The Silent Power of Creation

Neutrality, as Neville might describe it, is the art of detaching from the outer world to wield mastery over the inner. It's not the absence of emotion but the mastery of emotional equilibrium. It is the zero-point field within us, where we observe reality without judgement or reaction.

Imagine walking along a trail in the Nilgiris and coming across a rickety wooden bridge. You hesitate, unsure whether it's safe to cross. Fear creeps in, yet excitement beckons you to the other side. Neutrality is when you stop, breathe, and observe the situation without judgement. Instead of reacting impulsively, you choose your response deliberately.

Neville would call this the "silence before creation." He taught that every emotional reaction ripples through the quantum field, shaping the next reality we experience. Holding neutrality creates a blank canvas where conscious choices replace reactive patterns. Neutrality is not passivity but active creation. It's stepping back from chaos and painting your masterpiece with intention.

Fractal Time: The Infinite Now

As the Nilgiris remind us of their timeless beauty, time is an illusion—a fractal construct of infinite moments. Each moment contains the seeds of all possibilities. When we release the grip of linear thinking, we unlock the ability to shift seamlessly between these fractal moments, crafting new timelines.

Neville's concept of revision aligns perfectly with this idea. He taught that when we recall a memory, we're not merely reliving it but recreating it in the present. Imagine sitting on a hillside in the Nilgiris, watching the sun dip below the horizon. As the light fades, it feels like a memory, yet it's alive in the present moment. This interplay between past, present, and future is fractal time in motion.

For instance, if you remember an argument with a loved one, you can revise it. Replace anger with understanding and frustration with forgiveness. By doing so, you heal the past and alter the trajectory of future interactions, creating a reality infused with harmony.

Merging Realities: The Trinity Effect

Neville often emphasised the trinity of desire, feeling, and assumption. These three elements merge to create a new reality. In the quantum field, this process mirrors the merging of two parallel realities to form a third—a unified timeline born of intentional choices.

Picture yourself standing at a crossroads in life. One path offers stability, the other passion and risk. By embodying the feeling of excitement and success, you can merge the best aspects of both timelines. The trinity effect allows you to craft a new reality where stability supports your passions, and passion fuels your growth.

Neville echoes this process: "Assume the feeling of your wish fulfilled and continue feeling it until it becomes your reality." The frequency you hold determines which elements of your chosen timelines merge, shaping your experience.

Overwriting Negative Patterns

Negativity is a dense energy that anchors us to lower-frequency realities. We can transcend it not through resistance but through conscious replacement. Neville taught that what we resist persists, but what we assume transforms.

Consider standing amidst the Nilgiri tea gardens, feeling overwhelmed by self-doubt. Instead of suppressing it, you pause and layer over it with gratitude for the beauty surrounding you, love for yourself, and joy for the present moment. This 'vibrational layering' neutralises negativity, allowing you to shift effortlessly into a higher state of being.

Neville might describe this as living in the "feeling of the wish fulfilled." By consciously choosing higher vibrations, you overwrite the patterns that no longer serve you.

Practical Applications: A Day in the Quantum Field

Let's translate these ideas into a practical scenario. Imagine you're preparing to speak at a community event in the Nilgiris. As time approaches, anxiety creeps in, threatening to derail your confidence. Here's how you apply these concepts:

1. **Pause and Observe:** Acknowledge the anxiety without judgement. It's just a frequency, not your identity.
2. **Engage Neutrality:** Breathe deeply, grounding yourself in the present moment. Visualise the lush hills around you, steady and unshaken.

3. **Choose Your Frequency:** Close your eyes and feel the success of your talk. Hear the applause, see the smiling faces. Let the feeling wash over you.
4. **Reinforce the Vibration:** Repeat affirmations like, "I am calm, confident, and impactful," infusing them with genuine emotion.

Following these steps aligns with a reality where the event unfolds smoothly, guided by your intentional vibration.

Quantum Creation as a Way of Life

Living as a quantum creator is not about perfection but persistence. Each moment offers a chance to recalibrate and align your thoughts, emotions, and actions with your desired reality. Neville often said, "Your imagination is God's workshop." This workshop is open every moment, inviting you to craft your masterpiece.

The Nilgiris, with their changing seasons and steadfast presence, remind us that growth is a natural rhythm. There will be setbacks, doubts, and challenges, but even these are opportunities. They reveal where we are out of alignment, offering us the chance to course-correct.

So, step into your role as a quantum artist, painting your reality with neutrality, grace, and intentionality. Let the hills of the Nilgiris inspire you to live harmoniously, embracing the infinite possibilities of the quantum field. Every moment is a masterpiece waiting to be created—crafted by the silent power of your imagination.

CHAPTER 22

The Geometry of Giving: Unlocking Abundance Beyond Limits

The Nilgiris lay bathed in soft morning light, the peaks stretching skyward like aspirations taking form. From the tiny veranda of my guesthouse, I gazed over tea plantations cascading down hillsides like green waves. A koel called out in the distance, its song piercing the tranquil air. Nature seemed to give effortlessly, asking for nothing yet thriving in harmony.

As the mist cleared this morning, I reflected on an idea as expansive as these hills: the geometry of giving. Neville Goddard's insights came to mind. "The world is a mirror, forever reflecting what you are doing within yourself," he said. In a quantum reality where every action creates ripples of manifestation, giving unconditionally is not just an act of nobility but the cornerstone of true abundance.

The Structure of the Quantum Field

Neville described the quantum field as a vast hologram shaped by our thoughts, emotions, and actions—a mirror reflecting the intentions and vibrations we project into it. When aligned with love and authenticity, each act of giving enriches the field, creating ripples of abundance that return in unexpected ways.

Think of the tea leaves steeping in hot water, giving their flavour without expectation. Their essence transforms the water, just as our giving transforms the quantum field. Neville often emphasised the importance of assuming the feeling of the wish fulfilled. When you give from a place of abundance—assuming you already have enough—you align with the highest frequencies, inviting the universe to mirror that abundance.

The Trap of Conditions

During one of my walks through a Nilgiri village, I met Raghavan, a retired carpenter. He shared how, in his youth, he worked tirelessly, believing that hard work alone would bring him wealth. Yet, no matter how much he toiled, abundance always eluded him.

Neville's teachings illuminated the core of Raghavan's struggle. "You cannot serve two masters," Neville would say, referring to the conflict between giving freely and attaching conditions to it. Raghavan's mindset was transactional—his actions came with strings attached, narrowing the flow of abundance.

Everything changed when Raghavan shifted his perspective, offering his skills freely and trusting in the universe. He began to repair furniture for neighbours, accepting only what they could offer in return. Word of his generosity spread, and soon, opportunities came pouring in. A wealthy plantation owner commissioned him for a large project, securing his financial future. This transformation mirrored Neville's principle: "Give, and it shall be given unto you."

Giving and the Heart Frequency

The highest vibration in the quantum field is love, and its most authentic expression is through the act of giving. Neville often said, "Feeling is the secret," underscoring the power of emotions in shaping reality. When we give from the heart, without expectation, we align with the purest frequencies of the quantum field.

The Nilgiris themselves exemplify this principle. The monsoon rains nourish the tea gardens, the flowers feed the bees, and the eucalyptus trees offer shade. None of these elements demand anything in return, yet their giving sustains an ecosystem of abundance.

When we give from this heart-centred space, we become channels for the flow of infinite resources. The quantum field responds to the frequency of our emotions, amplifying and reflecting them as material abundance.

Gratitude: The Engine of Abundance

Just as giving aligns us with the quantum field, gratitude sustains that alignment. Neville taught that gratitude is the highest acknowledgement of abundance. To feel grateful even before receiving is to assume the wish is fulfilled.

Rukmini, a young artist I met at a roadside tea stall, embodied this truth. Struggling financially, she dreamt of hosting an art exhibition but had no means to achieve it. One day, she decided to sketch portraits of villagers, offering them as gifts. In return, she received meals, shelter, and connections. These acts of giving, infused with gratitude, eventually led to her first exhibition, which was a resounding success.

Neville's words resonate here: "Be thankful for what you have, and you will have more." Gratitude amplifies the vibrations of abundance, inviting the quantum field to deliver more of the same.

Reset: The Art of Letting Go

The Nilgiris are known for their sudden storms, where rain sweeps across the hills, cleansing the air and renewing the land. This natural rhythm reflects the power of resetting—letting go of old patterns to make way for new possibilities.

Neville's concept of 'revision' aligns beautifully with this idea. He believed that by mentally rewriting an event as you wish it to be, you could dissolve resistance and create a new reality. Resetting is not about abandoning your goals but releasing attachment to specific outcomes. It's an act of trust—trust in the universe's infinite intelligence to orchestrate events for your highest good.

When you give without clinging to expectations, you create space for the quantum field to surprise you. Letting go clears the way for synchronicities and manifestations far beyond what you could have imagined.

The Law of Abundance in Practice

Neville often said, "You do not attract that which you want, but that which you are." In quantum reality, abundance is not measured by material wealth but by the flow of resources, opportunities, and connections. The key to unlocking this flow lies in removing the barriers we impose—conditions, fears, and doubts—and embracing the freedom of unconditional giving.

The lesson is evident in the Nilgiris, where nature thrives on interdependence. When you give generously and align with gratitude, the quantum field mirrors your generosity, creating a cycle of abundance that benefits everyone.

Living the Geometry of Giving

As the sun sets behind the Nilgiris, casting a golden glow over the hills, I reflect on its interconnectedness. Every tree, river, and bird contributes to the greater whole, embodying the essence of abundance.

By embracing Neville Goddard's principles—the feeling of the wish fulfilled, the power of gratitude, and the art of letting go—we can live in harmony with the geometry of giving. Let the Nilgiris remind us that the greatest abundance comes not from clinging but from trusting, not from holding back but from giving freely.

Ultimately, our lives, like these hills, are shaped by the ripples we create. So give generously, live gratefully, and trust that the universe will always return your gifts in more beautiful ways than you can imagine.

CHAPTER 23

Becoming the Quantum Magnet: Collapsing the Chase

It was one of those still nights when the air seemed thick with mystery, the kind of night that invites introspection. Mist shrouded the hills outside my window, and the moonlight gently whispered through the clouds from the cosmos. The universe seemed poised to reveal a great secret, and a peculiar energy filled the air.

And so, I turned inward, seeking not the answers to life's pressing questions but the questions themselves. Why do we chase so much? Why do we believe the things we desire are running away from us? The very act of pursuit, I realised, affirms the absence of what we seek. We chase only the things we think we lack.

But what if we stopped chasing? What if we became the person who naturally attracts what we desire instead? What if we, like the mysterious quantum particles that physicists speak of, existed in multiple states—our present selves coexisting with the fulfilled versions of us?

This thought stirred me, like the rustling leaves outside my window. In the language of quantum mechanics, they call it superposition—the idea that particles don't commit to a single state until observed. Could it be that our lives are much the same? By focusing not on the chase but on the feeling of the outcome, we collapse the infinite possibilities into the one we wish to live.

The Magnetic Power of Being

Consider, if you will, a river. It doesn't chase the sea; it simply flows toward it, pulled by an invisible force. The river doesn't fret over how it will reach its destination. It knows. It trusts the gravity that guides it, the natural contours of the land that shape its path.

And yet, here we are, busy trying to bend the riverbanks of life, chasing outcomes with an urgency that often feels hollow.

Neville Goddard's philosophy echoes here, though I prefer to hear it in the whispers of the wind: **Assume the feeling of your wish fulfilled.** When you live as though your desire is already yours, you align with it. You magnetise it.

But there's a parallel in the quantum world, too—a phenomenon physicists call quantum entanglement. Two particles, no matter how far apart, remain connected. Change the state of one, and the other responds instantly. In the same way, your desires are not distant from you. Your desires, entangled with you, await your shift to react favourably.

The key lies not in doing but in being. When you become the person with what you desire, the world rearranges itself to meet your assumptions.

The Quiet Science of Imagination

Imagine, for a moment, that you are in a warm, sunlit meadow. The grass sways gently underfoot, and the air is rich with the scent of wildflowers. This scene exists only in your imagination, yet your body responds as you dwell in it. Your breathing slows. Your muscles relax. You feel calm, as though the meadow were real.

That's the power of your imagination. It's not merely an escape but a tool—a quantum workshop where we forge realities. Neville understood this, but so did neuroscientists and quantum physicists. When you vividly imagine an experience, your brain rewires as if the imagined scenario were real.

What is neuroplasticity? It is a concept as grounded in science as in mysticism.

Your imagination is a bridge to the quantum field of infinite possibilities. You alter your being by dwelling on your desire to be fulfilled. As the quantum field reflects what you project, the outer world aligns with the inner.

Living from the End

There's a story that illustrates this beautifully. A young artist struggling for recognition abandoned commission work and started painting as if she were already famous. She poured her joy and fulfilment into every brushstroke. She hung her paintings in her modest studio as though they belonged in galleries.

Months later, a collector found her work captivating. Soon, her paintings adorned the walls of some of the most prestigious galleries. What changed? She stopped chasing success and began living it.

In quantum terms, she tuned her inner frequency to match her desired outcome. Just as a radio needs to be tuned to the right station to receive a signal, she aligned herself with her desired reality.

The Role of Faith in the Quantum Field

Faith, I understand, is the willingness to persist in the unseen. It is the act of dwelling in the state of your wish fulfilled, even when the outer world shows no evidence.

Neville described this as feeling the reality of the unseen, but quantum physics offers a parallel: the observer effect. Particles behave differently when observed, as if responding to the act of attention. In much the same way, your life responds to the focus of your consciousness.

Faith is not unquestioning optimism. It is a deliberate observation, a conscious choice to dwell in possibility rather than limitation.

Letting It Come to You

And so, I ask you: What would happen if you stopped chasing? If you allowed the river to flow, trusting that the current will carry you? What if you turned your gaze inward, using your imagination not to escape but to create?

When you align with your desires—assuming the feeling of already being, having, or doing what you seek—the chase ends. Like a devoted partner, the universe conspires in your favour.

The night grew more profound, and the faint rustle of the eucalyptus leaves outside seemed to carry ancient secrets. Like a dream forgotten too soon, the mist cloaked the Nilgiri hills in a mysterious hush. From my window, the faint outline of tea plantations stretched into the distance, their symmetry disrupted only by the occasional lone oak. The world felt vast and intimate, as though the hills leaned closer, whispering truths to those who would listen.

And it was in this quiet embrace of the Nilgiris that I felt the futility of the chase dissolve. Like these undulating hills, life doesn't demand to be conquered; it asks only to be experienced.

The clouds drift lazily across the valleys, never rushing to reach the peaks. The koel sings without fretting about who might listen. Even the distant train, its whistle slicing through the stillness, winds gently along the slopes, trusting the rails beneath.

As the first light of dawn caressed the blue-hued mountains, I felt a quiet knowing settle within me. This was the lesson of the Nilgiris, written in every leaf and every ripple of the mountain streams: **Be still and let the universe come to you.** The hills do not chase the clouds, yet the rain comes. The trees do not race toward the heavens, yet they grow tall. So, too, must we align, trust, and allow.

In this timeless moment, I understood what it meant to become a magnet—embodying what we seek without chasing or grasping. The hills remained silent, but their message was clear: The universe always aligns if we let it.

CHAPTER 24

Quantum Creation: Teleportation – The Journey Without a Journey

The evening air was cool and crisp, enticing you to linger a little longer on your veranda, gazing at the world painted in twilight hues. The Eucalyptus trees stood sentinel-like against the fading light, their branches swaying gently in the breeze as if whispering secrets to the heavens. In such a moment, I pondered something extraordinary: the idea of travelling without ever moving, of collapsing the great expanse of space into a singular point. Teleportation—no longer the stuff of science fiction, but a reality that, like the gentle rustle of the Eucalyptus leaves, is slowly but steadily making itself heard.

There's magic in the very concept of teleportation, a sort of alchemy that transforms the impossible into the inevitable. But it is not magic. It is quantum mechanics, that enigmatic realm where particles dance to rules that defy common sense yet shape our universe. As science understands it, teleportation is not about physically transporting objects from one place to another. Instead, it's about transmitting their essence—their quantum state—to another location instantaneously, without crossing the space between them.

The Science of Leaping Across Space

Teleportation begins with quantum entanglement, that marvellously mysterious connection between particles that allows them to share information instantaneously, no matter the distance. Imagine two leaves on a Eucalyptus tree, trembling in perfect harmony despite being on opposite branches. Tug at one, and the other responds in kind, as though the invisible thread of the universe ties them together. This is entanglement—a phenomenon Einstein called "spooky action at a distance."

In laboratories around the world, scientists have already demonstrated quantum teleportation. They've taken particles of light—photons—and, using the magic of entanglement, transferred their quantum states to other photons across a distance. It's a delicate, precise process, like threading a needle in a gale, yet it has been done. What once seemed confined to the fantastical worlds of *Star Trek* has now taken its first tentative steps into our reality.

But this is not where the story ends; it is merely where it begins. If we can teleport the quantum state of a single particle, who's to say what might come next? Could we one day teleport molecules? Objects? Even ourselves? The path from particle to person may seem unimaginably vast, but so once did the idea of splitting the atom or landing on the moon.

The Journey Without a Journey

I often think of the journeys we take in life that shape us, transform us, and lead us to places we never thought we'd reach. Teleportation, in many ways, represents the ultimate journey without a journey. It's not about the steps you take or the miles you travel; it's about the destination simply arriving to meet you. The idea feels like something out of a dream, yet here we are, inching closer to making it real.

Imagine a world where distance becomes irrelevant. A world where you could step from one continent to another in the blink of an eye, not by traversing roads or skies but by collapsing the space between. It's not just a technological revolution; it's a profound shift in understanding the fabric of existence. Distance, after all, is a construct—a way for our minds to organise the world. Teleportation challenges that notion, suggesting that perhaps the universe is far more connected than we ever dared to believe.

Bridging the Imagination

Teleportation isn't just about particles or scientific breakthroughs; it's about what those breakthroughs mean for us as human beings. Neville Goddard often spoke of the power of imagination, of seeing the end before the beginning and living as though your desires have already been granted. In a way, teleportation mirrors this philosophy. It collapses the distance between where we are and where we wish to be, bringing the desired reality into being with startling immediacy.

Take, for example, the simple act of imagining a place you long to visit. Perhaps it's a sunlit meadow, a bustling city square, or even a loved one's embrace. In your mind's eye, you are already there. You feel the sun's warmth, hear the laughter of distant voices, or sense the comfort of that familiar touch. This act of mental teleportation isn't so different from what scientists are attempting to achieve on a quantum level. It is the mind's way of bridging the gaps that the body cannot.

The Future Unfolding

As I sit beneath the Eucalyptus trees, watching the first stars appear in the evening sky, I cannot help but wonder about the future. What might a world shaped by teleportation look like? Will it be a world of instant travel, where the boundaries of geography and time dissolve like mist in the morning sun? Or will it be something even more profound—a world where the very nature of connection itself is redefined?

Already, whispers of such a future are beginning to take shape. Scientists speak of using quantum teleportation to create un-hackable communication networks, where information travels not through wires or airwaves but through the very fabric of the universe. Others envision teleportation transforming industries—medicine, logistics, space exploration—on a scale we can scarcely imagine.

And yet, for all its promise, teleportation also reminds us of the importance of the journey. There is beauty in travelling, the slow unfolding of landscapes and the chance encounters along the way. Teleportation may one day make such journeys unnecessary, but it will never diminish their value. Even as we leap across the stars, the steps we take, however small, define us.

Closing Thoughts

The night deepens, and the Eucalyptus trees stand silhouetted against a sky now scattered with stars. Somewhere out there, in the vastness of the cosmos, particles are dancing, entangling, and teleporting, weaving the threads of a story that we are only beginning to understand. It is a story of connection, of imagination, and of the infinite potential that lies within us all.

Teleportation is more than a scientific achievement; it is a testament to humanity's curiosity and creativity. It invites us to dream bigger, question the limits of what we know, and embrace the unknown with open hearts and minds. It is, in every sense, a journey without a journey—and one that has only just begun.

CHAPTER 25

The Alchemy of Energy Quantum Creation: Consciousness as the Gateway to Infinite Potential

Evening falls gently in the hills, and the Eucalyptus trees sway meditatively. Their rustling leaves seem to murmur secrets, like the universe whispering tales of quantum mysteries and human potential. Seated with my notebook, I watch as the last rays of sunlight stretch across the horizon, and it occurs to me: there is no separation between the vastness of the cosmos and the intricate thoughts within us.

This chapter is about that connection—the role of consciousness in quantum creation. From healing the human body to advancing material sciences, consciousness emerges as a force that transcends the physical, bridging the realms of science and spirituality. Neville Goddard's teachings—that imagination and belief shape reality— intertwine seamlessly with these scientific insights, inviting us to explore the limitless possibilities of human thought.

Consciousness as a Diagnostic System

Traditional medicine relies on tools and technologies to probe the physical body. But what if there was another way? What if consciousness itself could serve as a diagnostic system, bypassing machines and relying instead on the intuitive awareness of the human mind?

Take, for example, the phenomenon of intuitive diagnostics—where practitioners, through heightened states of awareness, can detect illnesses that even advanced medical tools might overlook. Pattern recognition beyond data lies at the heart of this capability. While machines depend on measurable inputs, the mind can perceive subtle energetic patterns—shifts in mood, vitality, or the unspoken signals of the body.

Neville Goddard often emphasised the power of belief in healing. He would ask: What if the mind could "assume" a state of health so vividly that the body could not help but align with it? Today, this idea is gaining traction in metaphysical circles and cutting-edge medical research. Meditation, lucid dreaming, and altered states of consciousness have shown remarkable potential in enhancing intuitive capabilities. In these states, individuals access a deeper layer of reality—a universal consciousness field where answers to the body's mysteries reside.

Case Study: The Healing Mind

A renowned study involved patients visualising their immune cells attacking cancer cells. Over time, those who practised visualisation demonstrated significant improvements compared to the control group. It wasn't the imagery alone but the belief and focus—the conscious intent—that unlocked the body's innate ability to heal. This is not mere pseudoscience; it is quantum creation in action, where thought becomes a tangible force for transformation.

Reverse Engineering Advanced Materials

The universe is a storehouse of secrets, and some of its greatest treasures lie hidden in the materials that form its fabric. Reverse engineering advanced materials—especially those inspired by extraterrestrial technologies—requires a blend of quantum insight and human intuition.

Picture a scientist standing before a crystalline metallic structure. Traditional tools might analyse its composition, but only consciousness can perceive its vibrational signature—the unique energetic blueprint that defines its properties. By interfacing with the quantum field, researchers have uncovered materials with extraordinary capabilities, from superconductors to lightweight alloys with unprecedented strength.

Neville Goddard's philosophy finds a parallel here. He often spoke of the unseen forces that shape visible reality. In reverse engineering, it is not just the microscope or the data that matters—it is the act of focused attention, the willingness to "see" beyond the obvious, that leads to breakthroughs.

Practical Outcomes

Consciousness-based insights have already revolutionised material sciences. Imagine a spacecraft coated in heat-resistant materials inspired by these discoveries, or energy grids powered by superconductors that lose no energy to resistance. These are not distant dreams but active projects, blending the intuitive capabilities of the human mind with the precision of quantum computation.

Consciousness-Based Communication Systems

In a world driven by connection, instantaneous communication is both thrilling and necessary. Traditional systems rely on electromagnetic waves constrained by the speed of light. But consciousness-based communication—leveraging quantum entanglement and non-locality—offers a glimpse into a future without such limitations.

Non-Local Interaction

Quantum entanglement defies distance. When particles are entangled, a change in one instantly affects the other, regardless of separation. Now, imagine extending this principle to human thought. Communication could transcend distance and time if consciousness is interconnected through a universal field.

Advanced systems are already exploring this. Devices interfacing with human thought could enable real-time global connectivity, where the mind acts as both sender and receiver. Neville Goddard might have called this the "assumption" of connection—the act of believing so firmly in unity that it manifests instantaneously.

Thought as a Computational Signal

Consider this: What if your thoughts could directly control machines? In these systems, consciousness becomes the signal, transmitting intent and emotion as data. The implications are profound. From controlling spacecraft to managing global crises, thought-based systems could revolutionise how we interact with technology.

Sub-Light-Speed Computation and the Quantum Mind

The laws of physics bind traditional computation, but sub-light-speed computation, powered by consciousness, operates in an entirely different paradigm. Here, the mind integrates with quantum systems, creating a symbiosis of intuition and logic.

Holographic Processing

Consciousness doesn't work linearly. It operates holographically, where every fragment contains the whole. This principle, mirrored in quantum systems, enables instantaneous access to vast information. Imagine a computer that doesn't process step by step but instead "knows" the answer by accessing the quantum field. This is the future of computation.

Beyond the Brain

The power of consciousness-based computation isn't confined to the human mind. It extends to external systems, creating a partnership between thought and machine. Neville Goddard's teachings resonate here—he believed the mind could project its reality outward, shaping the world. Consciousness-based computation furthers this, blending inner intention with outer action.

Challenges and Ethical Considerations

Every leap in human capability brings challenges, and quantum creation is no exception. The integration of consciousness into systems raises critical questions:

Reliability: Consciousness, though powerful, is influenced by emotions, biases, and distractions. How do we ensure consistent results?

Privacy: If thoughts can transmit data, who safeguards the sanctity of the mind?

Equality: Access to these advancements must be universal, or they risk widening societal divides.

Neville Goddard emphasised the ethical responsibility of creators. What we imagine, we manifest—and with great power comes great accountability. The same applies here. As humanity steps into this new era, it must do so with compassion and foresight.

The Future of Consciousness and Quantum Creation

I am filled with quiet awe as the stars emerge above the Eucalyptus grove. Once dismissed as intangible, consciousness is the most profound force. It is the key to unlocking the mysteries of the universe and our potential.

Neville Goddard believed in living "as if"—acting as though your dreams were already confirmed. With its blending of consciousness and science, Quantum Creation gives us the tools to do just that. It invites us to imagine boldly, create responsibly, and embrace the infinite possibilities of existence.

This journey is only beginning. As the Eucalyptus leaves whisper their quiet symphony, I am reminded that the most incredible adventures often start with a single thought. Let us think wisely, act with intention, and create a world worthy of the boundless potential within us all.

Conclusion – The Future of Quantum Manifestation

As the sun dipped below the horizon, casting a golden glow over the landscape, the day transitioned into a quiet evening—a moment where the visible world meets the unseen, where the boundaries between day and night blur, much like the boundaries between science and spirituality. In this serene twilight, we stand at the culmination of our journey through quantum physics and Neville Goddard's profound insights. Here, in this space of transition and possibility, we reflect on the incredible power of thought, the role of consciousness, and the interconnectedness of all things.

Recap of Key Concepts

Throughout this book, we have traversed the fascinating world of quantum mechanics, uncovering its foundational principles and discovering how they resonate deeply with the metaphysical teachings of Neville Goddard. Let's revisit the fundamental concepts as they form the bedrock of our understanding of quantum creation.

Wave-Particle Duality

One of the most intriguing aspects of quantum mechanics is that particles can exhibit wave-like and particle-like properties, depending on how we observe them, mirroring Goddard's teaching that our thoughts and feelings are not fixed but fluid and capable of shaping our reality. Just as light can be both a wave and a particle, our consciousness can hold multiple potential realities simultaneously, allowing us to shape our world through the power of our imagination.

Superposition

The concept of superposition, where particles exist in multiple states until observed, aligns beautifully with the Law of Assumption. Goddard teaches that by assuming the feeling of the wish fulfilled, we hold the potential for multiple outcomes waiting for realisation within us. In the quantum world, this means that reality is not a single, predetermined path but a spectrum of possibilities we can influence with our thoughts and intentions.

Entanglement

Perhaps one of the most profound concepts in quantum mechanics is entanglement—the idea that particles, once connected, remain linked regardless of distance. This phenomenon echoes Goddard's belief in the interconnectedness of all things. Our thoughts, emotions, and intentions are not isolated; they ripple into the universe, influencing distant outcomes and connecting us to a more extensive web of existence. This interconnectedness reinforces the idea that we are all part of a greater whole, where our actions can have far-reaching effects.

The Observer Effect

The observer effect, which demonstrates that observation can influence the outcome of a quantum event, underscores the power of consciousness in shaping reality. Goddard's teachings emphasise the importance of focusing on our desired reality and assuming it to be true. In doing so, we act as the observers who influence the quantum field, manifesting our desires.

Quantum Consciousness

Emerging research suggests that human consciousness may be a fundamental aspect of the universe intertwined with the fabric of reality.

This concept aligns with Goddard's assertion that our imagination and beliefs are not just passive reflections of the world but active tools for creation. The mind interacts with the quantum field through its focused intent, influencing the probabilities and possibilities that shape our reality.

The Future of Quantum Manifestation

The potential of quantum manifestation extends beyond personal success stories. It holds the promise of transforming not only individual lives but also the collective consciousness of humanity. By understanding and harnessing the principles of quantum mechanics and metaphysics, we can create a more harmonious, abundant, and interconnected world.

Collective Consciousness

The idea that our collective thoughts and intentions can influence global events is gaining traction. Initiatives such as the Global Consciousness Project, which measures the impact of collective human consciousness on random number generators, suggest that our collective focus can create measurable changes in the physical world. As more people become aware of their power to influence reality, the collective consciousness can shift towards creating a world grounded in compassion, unity, and shared purpose.

Healing and Well-Being

Quantum principles are finding application in medicine and healing. Techniques such as energy healing, quantum touch, and intention-based therapies are gaining recognition for their potential to promote physical and emotional well-being. By aligning our thoughts and intentions with health and vitality, we can support the body's natural healing processes. These approaches offer a glimpse into how we can harness quantum consciousness to improve health outcomes, fostering a deeper understanding of the mind-body connection.

Environmental Impact

The interconnectedness of all things, as highlighted by quantum entanglement, underscores the importance of our relationship with the environment. By cultivating a sense of unity and responsibility, we can contribute to the healing and preservation of our planet. Initiatives focused on sustainable living, conservation, and environmental consciousness exemplify how quantum principles can inspire positive change. Just as quantum physics reveals the interconnectedness of all things, our actions, thoughts, and intentions are intertwined with the health of our planet.

A Call to Action

Embrace Your Power

Recognise and embrace your innate power to create and manifest. Trust in your ability to influence the quantum field and realise your desires. Your thoughts, feelings, and beliefs are potent tools for creation. By acknowledging your role as a co-creator in the universe, you empower yourself to shape your life and the world around you consciously.

Share Your Journey

Inspire others by sharing your experiences and insights. Your journey can serve as a beacon of hope and possibility for those seeking to understand and harness the power of quantum manifestation. By sharing your story, you contribute to humanity's collective awakening and empowerment. Like the ripple effect in quantum entanglement, your actions and experiences can influence and inspire those around you, creating a wave of positive change.

Create a Better World

Use the principles of quantum manifestation to create a better world for yourself and others. Cultivate compassion, kindness, and gratitude in your interactions. Focus on positive intentions and envision a future filled with harmony, abundance, and interconnectedness.

The principles of quantum mechanics and metaphysics are not just abstract concepts but tools for creating a more compassionate, sustainable, and interconnected world.

Final Thoughts

As we conclude this exploration of quantum creation, I leave you with a sense of wonder and possibility. The intersection of quantum physics and Neville Goddard's teachings offers a profound and transformative perspective on the nature of reality and the power of the mind. By embracing these principles, you embark on a journey of self-discovery, empowerment, and creation.

The future of quantum manifestation is bright, filled with endless possibilities and opportunities for growth. As you continue to explore, experiment, and manifest, remember that you are a co-creator of your reality, capable of shaping the world in ways that align with your highest aspirations.

May your journey be filled with wonder, inspiration, and the realisation of your deepest desires. The power to create is within you, and the universe is your canvas. Paint your reality with the colours of your dreams, and watch as the quantum field responds to your intentions. The principles of quantum mechanics and metaphysics are not just theories—they are the keys to unlocking the boundless potential within each of us.

Thank you for joining me on this journey. The adventure of quantum creation has only just begun, and the possibilities are limitless. Embrace your power, trust in the process, and manifest the life you envision. The future is yours to create.

Author Bio

Dr. K. Jayanth Murali is a quantum alchemist in a world of probabilities, blending the deterministic rigor of science with the fluid wonder of metaphysics. A retired Director General of Police, celebrated author, and lifelong seeker, Dr. Murali's journey defies the boundaries of disciplines and dimensions. He is not merely a scientist or a philosopher; he is the bridge where knowledge meets imagination and where thought becomes reality.

Born in the shadow of the historic Golconda Fort, he grew up surrounded by stories of resilience and reinvention. From decoding the mysteries of microbial life during his academic years to pioneering the use of blockchain in policing, Dr. Murali's life has been a series of quantum leaps. With over three decades in law enforcement, he redefined what it means to lead with vision, integrating emerging technologies to bring justice and innovation into harmony.

Beyond his professional achievements, Dr. Murali is a relentless explorer of human potential. His literary works traverse the spectrum from peak psychology to the metaphysics of manifestation, each reflecting his quest to understand the mechanics of reality. His latest, The Art of Peak Performance: Hacking the Body and Mind For Peak Success, merges cutting-edge science with timeless wisdom to illuminate the pathways of human excellence.

An endurance athlete who has completed over 50 marathons, Dr. Murali views running as a form of quantum alignment—a meditation in motion where mind, body, and spirit converge. His advocacy for organ donation through marathon initiatives embodies his belief in entangling personal passions with universal good.

Whether painting vivid landscapes, cultivating life on his farm, or engineering solutions during the pandemic, Dr. Murali's pursuits reflect his commitment to serving both the tangible and intangible. For him, the universe is not a series of isolated events but an interconnected field where every action ripples through space and time.

At his core, he is Jayanth—the husband to Dr. Jayanthi, father to Anisha and Anussha, and a human being who cherishes the quantum entanglement of love, laughter, and legacy. His life is a testament to the idea that in the quantum field of existence, the observer and the observed are forever intertwined.

Discover more about Dr. Murali's explorations into the quantum mechanics of life at www.jayanthmurali.com.

www.ingramcontent.com/pod-product-compliance
Lightning Source LLC
Chambersburg PA
CBHW060526160726

47991CB00001B/197